TK2921
M35

Mantell, C. L.
    Batteries and energy systems.

PROPERTY OF
NEW ENGLAND TECHNICAL INSTITUTE

Batteries and Energy Systems

Batteries and Energy Systems

C. L. Mantell, Ph.D.

Consulting Engineer
AUTHOR: *Electrochemical Engineering, Carbon and Graphite Handbook*
EDITOR: *Engineering Materials Handbook*

McGraw-Hill Book Company

New York   St. Louis   San Francisco   Düsseldorf   London
Mexico   Panama   Sydney   Toronto

BATTERIES AND ENERGY SYSTEMS

Copyright © 1970 by McGraw-Hill, Inc. All Rights Reserved.
Printed in the United States of America. No part of this
publication may be reproduced, stored in a retrieval system,
or transmitted, in any form or by any means, electronic,
mechanical, photocopying, recording, or otherwise, without
the prior written permission of the publisher.
Library of Congress Catalog Card Number 70-107448

40030

1234567890 MAMM 7543210

Preface

Military demand for packaged power for communication, radiosondes, meteorological units, underwater propulsion, space vehicles, along with single-cycle convenience units; cordless appliances, which are rechargeable; utility items as represented by electric watches, hearing aids, flashlights, pocket radios, electric clocks, photoflash devices; emergency lighting, industrial lanterns, buoys, transistor appliances, construction flashers, distress signalling, rescue devices; civilian demand for rotating toothbrushes, knives, dictating machines, tape recorders, alarm systems, telemetering, hand tools, movie cameras, and toys; power packs, lawn and garden tools; as well as a host of other applications; has spawned myriads of batteries of all sorts and descriptions in all the nations of the world and has developed a multibillion dollar business. In this volume, dry cells, obsolete units, mercury cells, silver cells, air cells, fuel cells, storage batteries, regeneration systems, standards and auxiliaries, are treated as components, competitive to be sure, of a coherent whole, the energy systems which convert the activity of chemical reactions to convenient, but highly useful electric energy at a time, place, and under circumstances where desired or needed.

Thanks are owing to the battery companies, Bright Star, Burgess Battery Division of Clevite Corporation, Delco-Remy Division of General Motors, Eagle-Picher Industries, ESB Incorporated, and its divisions, Eveready of Union Carbide, Globe-Union Inc., Gulton Industries, Inc., Le Carbone, Marathon, Ray-O-Vac, RCA, and Whittaker Corporation.

Primary battery sales are projected to increase from $260 million in 1966 to $605 million by 1980. Sales of the zinc-carbon battery will double by 1980 and reach $383 million, but this will represent a declining share of all primary batteries. Alkaline-manganese oxide and others will grow the fastest.

Secondary battery sales are projected to increase from $700 million in 1968 to $1.2 billion by 1980. Lead-acid batteries account for almost 90% of the total with automotive types predominant. Automotive batteries will continue to increase because of expanded auto and truck production and increased replacement needs. However, the fastest growing types will be lead-acid batteries for industrial trucks, nickel-cadmium batteries for cordless appliances and other consumer use, and specialty batteries such as sealed lead-acid, silver-zinc, silver-cadmium, nickel-iron and nickel-zinc.

Fuel-cell use has been primarily in the Gemini and Apollo programs, but future applications will include military and commercial terrestrial uses such as communications, electrical needs for housing and trailers, and as a replacement for the internal combustion engine in items such as boats, mowers, and tractors.

The proponents of electric-auto development hope for market acceleration.

Because of air pollution problems, a number of battery, fuel-cell, and hybrid systems are now under development.

Despite a rapidly expanding research and development effort on solar cells, fuel cells, and other energy conversion systems, older systems will dominate the direct-power market through 1980. Sales of batteries have increased from $510 million in average 1957–1959 to $867 million in 1966 and are projected to exceed $1.8 billion by 1980. The markets for fuel cells, solar cells, and other energy conversion systems, have been held back by cost and other factors and have been confined largely to space uses. It appears that only fuel cells will make any appreciable gains before 1980, since solar-cell growth will be in the light sensing uses.

The battery business, historically considered electrochemical, combines the activities of the chemist, the chemical engineer, the metallurgist, the materials specialist, the electrical engineer, the electronic designer, the mechanical coordinator, the space specialist, the environmental engineer, the instrumental analyst, and a variety of coordinating standardization organizations, testing agencies, as well as the application originator for expanded and new uses. All these have individual viewpoints which cooperate to serve the diversified needs and desires of our industries, in addition to our daily lives in the electronic and space age.

The valued cooperation of Mr. Frank M. de Santa, who prepared the manuscript, is appreciated.

C. L. Mantell

Contents

PREFACE *v*

1. *History and the Galvanic Concept*   *1*
   Technical and Patent Literature

2. *The Voltage Concept—Standard Cells*   *8*
   Preparation. Properties. Reference Electrodes

3. *The Current-producing Cells and Batteries*   *26*
   Comparative and General. Economics

4. *The Work Horse—Dry Cells—the Zinc, Ammonium Chloride/ Manganese Dioxide/Carbon System*   *33*
   Manufacture. Battery Service Life. Quality Control and Testing. Sizes and Application. Recharging Primary Batteries. Magnesium Batteries

5. *The Zinc-Alkali-Manganese Dioxide Primary Batteries*   *54*
   Alkaline Cells

Contents

6. **Air-depolarized Cells** — 59
   Porous and Adsorptive Carbon Electrodes. Competitive Types and Application

7. **Fuel Cells** — 64
   Types and Preparation. Fuels

8. **Mercury Cells** — 78
   Types and Listing. Preparation. Sizes and Applications

9. **The Silver Batteries** — 85
   Types. Manufacture. Sizes and Applications

10. **Water-activated Systems** — 97
    Magnesium-Silver Chloride. Magnesium-Cuprous Chloride. Sizes and Applications

11. **Obsolete and Historical Systems** — 102

12. **Reversible Systems—Secondary Cells** — 107

13. **Lead Secondary Cells** — 110
    Theory. Manufacture and Testing. Sizes and Applications. Dry-charged Batteries. Counter Cells. Lead-Calcium Cells

14. **Alkaline Secondary Cells** — 155
    Type and Theory. Alkaline-Manganese Dioxide

15. **The Nickel-Cadmium System** — 163
    Unsealed and Sealed Cells. Silver-Cadmium Cells

16. **Battery Charging: Theory and Practice** — 176

17. **Regenerative Electrochemical Systems** — 183

18. **Solar Cells and Related Systems** — 188

19. **Development and Specialized Application Cells** — 191

20. **Electric Cars and Batteries** — 201

21. **Selection of a Battery** — 207

INDEX 215

Symbols and Abbreviations

A_p, A_r	activities of reactants	asc	amperes per square centimeter
a	activity		
AAA	American Automobile Association	asf	amperes per square foot
		asi	amperes per square inch
AABM	Association of American Battery Manufacturers	ASTM	American Society for Testing and Materials
abs	absolute	atm	atmosphere
ac	alternating current	Ba	barium
ACS	American Chemical Society	Br	bromine
AEC	Atomic Energy Commission	°C	degrees Celsius (Centigrade)
Ag	silver		
ah	ampere-hours	c	concentration
Al	aluminum	Ca	calcium
amp	ampere	cc	cubic centimeters
amp-min	ampere-minute	Cd	cadmium
ANSI	American National Standards Institute, formerly ASA, USASI	cemf	counter electromotive force (cells)
		cgs	centimeter-gram-second system
As	arsenic		

xi

Symbols and Abbreviations

Cl	chlorine	kw	kilowatt
cm	centimeter	kwh	kilowatthour
Co	cobalt	l	liter
coul	coulomb	lb	pound
cp	candlepower	Li	lithium
CR	charge-retaining (battery)	M	metal
Cr	chromium	m	meter
Cu	copper	ma	milliampere
cu	cubic	mah	milliampere-hour
DB	double insulation	max	maximum
dc	direct current	Mg	magnesium
E	voltage or potential	Mn	manganese
e	electron	min	minute, minimum
E_c	counter emf	M.I.T.	Massachusetts Institute of Technology
E_s	electrode potential		
E_{sb}	basic potential	mks	meter-kilogram-second system
E_{so}	normal electrode potential		
emf	electromotive force	mm	millimeter
ESB	Electric Storage Battery, Inc.	mole %	mole percent
		mph	miles per hour
F	fluorine	msec	millisecond
F	free energy	mwsc	milliwatts per square centimeter
ΔF	change in free energy		
$\mathcal{F}$	Faraday constant	N	nitrogen
°F	degrees Fahrenheit	N	normal solution; number of Faradays
f	farad		
Fe	iron	n	negative
G	electric conductance	n	number; difference in valence
G	specific gravity (of electrolyte)		
		Na	sodium
g	gram	NASA	National Aeronautics and Space Administration
g	specific gravity (of water at cell temperature)		
		NBS	National Bureau of Standards
GM	General Motors		
H	hydrogen	NEDA	National Electronics Distributors Association
H	heat of reaction		
Hg	mercury	Ni	nickel
hr	hour(s)	O	oxygen
Hz	hertz (1 cycle per second)	OART	Oakland Army Terminal
I	electric current, iodine	ohm-cm	ohm-centimeter
in.	inch	oz	ounce
IRE	Institute of Radio Engineers	P	plastic
		p	positive
K	potassium	P	electric power, reactant, osmotic pressure
°K	degrees Kelvin		
k	dielectric constant	p	electrolytic solution pressure
kg	kilogram		

Pb	lead	Sb	antimony
pH	negative logarithm of the effective hydrogen-ion concentration	sec	second
		sq	square
		T	absolute temperature
p-n	positive-negative	TRAC	Thermally Regenerative Alloy Cell
ppm	parts per million		
psi	pounds per square inch	v	volt(s)
psig	pounds per square inch gage	w	watt(s)
Pt	platinum	wh	watthour
PVC	polyvinyl chloride	x, y	coefficients
%	percent	Zn	zinc
Q	electric quantity		
R	resistance, rubber		
R	gas constant		
R_i	internal resistance		
R_s	series resistance		
rpm	revolutions per minute		
S	sulfur		
SAE	Society of Automotive Engineers		

Greek symbols

γ	conductivity
μ	micro
ρ	electric resistivity

**Batteries and
Energy Systems**

chapter 1
History and the Galvanic Concept

Batteries were the first source of electric energy in the development of electrochemical processes and laws. "Static" electricity made by rubbing dry cloths against amber was known to the Greeks and Egyptians before the Christian era.

Willard F. M. Gray, of General Electric, Pittsfield, Massachusetts, made working replicas of supposedly 2,000-year-old wet cells. The B.C.-vintage batteries which served as Gray's models (made by the Parthians, who dominated the Baghdad region between 250 B.C. and A.D. 224) were thin sheet copper soldered into a cylinder 1.125 cm long and 2.6 cm in diameter—roughly the size of two flashlight batteries end-to-end. The solder was a 60/40 tin-lead alloy. The bottom of the cylinder was a crimped-in-copper disk insulated with a layer of asphalt (the "bitumen" that the Bible tells us Noah used to caulk the Ark). The top was closed with a one-hole asphalt stopper, through which projected the end of an iron rod. In order that the cell would stand upright, it was cemented into a small vase.

A difference in potential is set up when two unlike metals are dipped into an electrolyte. The copper-iron combination in these ancient batteries is the same as that which Luigi Galvani used in 1786, when he "discovered" the galvanic cell.

This evidence was discovered 20 years ago by Wilhelm König, a German archeologist, at the Iraq Museum. A small hill of Khujut Rabu'a, on the outskirts of Baghdad, was being dug away, and the remains of a Parthian town were revealed. The museum began scientific excavations, and the digging turned up a peculiar object that—to König—looked like a dry cell.

The archeologist learned that four similar vases, plus slender bronze and iron rods that looked like connecting wires, had been turned up in the ruins of a magician's hut in the ancient city of Seleucia, down the river. Back in the Berlin Museum he found the unassembled parts—copper cylinders, iron rods, asphalt stoppers—for what could be 10 more cells like the ones from Khujut Rabu'a.

König described his finds in a book on his 9 years in "the Lost Paradise" of Iraq. The accounts were spotted by science historian Willy Ley, who reported them in English and American journals. Willard Gray volunteered to test König's theories by duplicating one of the batteries and seeing whether it worked; Ley was able to supply him with dimensions, diagrams, and analyses of the metals. Gray's model is now in Pittsfield's Berkshire Museum.

In the period from 1798 to 1800, Alessandro Volta developed the voltaic pile of unlike metals in contact with an electrolyte, a succession of these in series as a voltaic source, larger areas of metal (or electrodes) for greater current, and the "crown of cups."

Early experimenters had suspected that there was a relationship between chemical and electrical phenomena. It remained for Volta to confirm this relationship. The original "voltaic pile" consisted of a series of zinc and silver disks separated from each other by a porous nonmetallic material and made electrically conductive by impregnation with salt water. This arrangement produced a voltage across each silver and zinc disk. Volta arranged these disks as shown in Fig. 1.

Another arrangement demonstrated by Volta was the crown of cups, a group of cups containing salt water, arranged in a circle, and connected to each other by conductors with terminating electrodes of zinc and silver. This arrangement is illustrated in Fig. 2. Neither the voltaic pile nor the crown of cups was a practical battery because of their bulk and awkward arrangement of cells.

Wallaston later built very large batteries for the Royal Institution in London. These were the power sources for the work of Davy and Faraday in the next 20 years. Batteries were the source of electrical energy until the development of the dynamo on an industrial scale long after Faraday had pointed out the inductive effects of the electromagnet.

When central stations were developed by Edison and Weston, it appeared that batteries would lose their importance.

A major advance in the evolution of the battery was the cell named for its inventor, J. F. Daniell. This unit incorporated a depolarizing agent

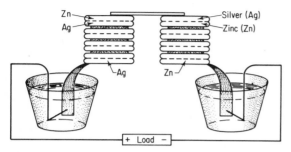

Fig. 1 Voltaic pile. (*Radio Corporation of America.*)

(a material which reduces the accumulation of hydrogen on the electrode) which aided in extending the life. The cell utilized a zinc negative electrode immersed in a dilute acid electrolyte (zinc sulfate + sulfuric acid) and a copper positive electrode immersed in a copper sulfate solution. Before the Civil War in the United States, Planté had developed the essentials of the lead–sulfuric acid storage cell, chargeable by steam-engine-driven dynamos.

In 1868 Georges Leclanché introduced a cell which was the forerunner of the present-day cell. Because of these chemical similarities, the dry

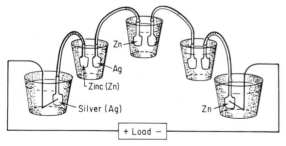

Fig. 2 Volta's crown of cups. (*Radio Corporation of America.*)

cell, called the zinc-carbon cell, is still referred to as a Leclanché-type cell. The Leclanché cell had the feature of employing only one liquid material, an ammonium chloride (sal ammoniac) solution which replaced the acid electrolyte used in earlier cells. The depolarizing solution was

TABLE 1 Battery Chronology

Year	Person	Description
1800	Volta	Unlike metals and electrolyte in contact—"voltaic piles," first power source
1800	Nicholson and Carlisle	Electrolysis of water into hydrogen and oxygen
1807	Humphry Davy	Electrolysis of caustic—sodium, potassium, etc., using voltaic "batteries"
1810	Humphry Davy	Electric arcs of carbon, powered by batteries
1816	Robert Hare	Calorimotor, battery for heating purposes
1827	Georg Simon Ohm	Ohm's law
1830	Michael Faraday	Electrochemical laws
1834	Michael Faraday	Inductive effects of the electromagnet
1836	Daniell	$Cu|CuSO_4|ZnSO_4|Zn$ chemical battery
1836	Elkington	Electroplating of silver
1839	Jacobi	Electrotyping of articles
1840	Spencer and Jordan, Wright	Electroplating solutions; essentials
1846	Bottger	Electrodeposition of iron
1849	Russell and Woolrich	Cadmium plating
1852	Bunsen	Fused-salt electrolysis; magnesium
1855	Bunsen	Fused-salt electrolysis; lithium
1859	Planté	Lead storage cell or battery
1868	Leclanché	$Zn|NH_4Cl|C$ battery
1871	Robinson	Train signaling by batteries
1874	T. A. Edison	Quadruplex telegraphy by batteries
1880	Niaudet	Book on batteries
1881	Faure	Pasted plates in lead-acid batteries
1882	Sellon	Antimony-lead alloy for grids
1882	Gladstone and Tribe	Double sulfate theory
1882–1886	....	Zinc–silver chloride batteries
1888	Gassner	"Dry" cell
1900	T. A. Edison	Nickel storage battery
1910	T. A. Edison	Commercial nickel-iron alkaline cell
1910	....	Flashlight cells D size
1910	....	Telephone dry cell
1912	....	Standard method of testing batteries
1917	....	First battery specifications
1924	....	Standard nomenclature for batteries
1928	Arendt	Book on storage batteries
1930	....	Industrial flash cells
1931	U.S. Government	Federal purchase specifications for batteries
1932	....	Hearing aid (CD) size commercial
1940	....	Underwater propulsion by batteries
1941	....	Radiosonde units
1941	....	Batteries for rockets and missiles
1942	....	World War II batteries
1943	Adams	Cuprous chloride battery
1945	Ruben	Mercury cell
1955	....	Alkaline cell commercial development
1959	....	Current American standards

replaced by a dry mix composed of manganese dioxide and carbon. Imbedded in the center of this mix was a carbon bar which served both as a current collector and as the positive electrode. An advantage of the Leclanché cell over the Daniell cell was its higher electromotive force (voltage). Although superior to the Daniell cell, the Leclanché cell was still restricted to laboratory and fixed installations because of its liquid content.

The first true dry cell was developed between 1886 and 1888 by Dr. Carl Gassner. This unit used a paste electrolyte composed of zinc oxide, sal ammoniac, and water. The zinc negative electrode was modified so that it also served as the container for the cell contents. The carbon rod was the positive electrode, located in the center of the battery. To prevent leakage and evaporation, the space between the carbon electrode and the zinc container was sealed at the top with plaster of paris. The result was a cell which was portable and adaptable to varying space requirements. Several could be connected to form batteries for higher voltage and/or current requirements.

Dr. Gassner's development made it practical to manufacture dry cells on a commercial scale. Commercial production of the Gassner cell began in the United States shortly after its discovery.

Battery chronology in Table 1 shows that many developments are more than a century old, but developments are still taking place at a rapid rate.

The battery was the first practical source of electrical energy developed in man's search for portable power sources. Although many other techniques have been developed for supplying electrical power, the battery, which converts chemical energy directly into electrical energy, is still the most widely used source of electrical power when portability is the prime requisite.

The development of semiconductor devices such as transistors, diodes, missiles, satellites, and a great variety of mobile equipment has imposed rigorous demands for power sources which are compact, dimensionally adaptable, able to operate over a wide temperature range, and highly dependable. That the battery meets these demands is proved by the enormous increase in battery use and continuous demands for new battery types.

Technical and Patent Literature

The literature on electrochemical energy systems or batteries is not only almost a century old in certain sections, but voluminous and extensive. In part it carries an air of mystery and mysticism. Each new developer promised minor miracles with unrestrained enthusiasm.

A literature search, limited to dry cell technology, of the Battery Branch of the Signal Corps Engineering Laboratory, Fort Monmouth, New Jersey in 1948 listed several thousand patents since the turn of the century.

The major lead–sulfuric acid secondary cell patents expired about the turn of the century. "Improvement" patents since that time have covered hardware, appurtenances, connectors, and containers.

It is interesting to note that the Adams 1943 patent for the magnesium–cuprous chloride water-activated battery was the only patent in the numerous ones litigated that has been upheld by the Supreme Court since 1945. There are several "improvement" patents. The alkaline Edison cell patents had expired before World War I. The Ruben 1945 mercury cell patents have expired and there are many mercury cell battery manufacturers.

Arendt's 1928 volume on storage batteries developed from lectures to engineering students at Columbia and to officers at the U.S. Submarine School over a period of years. He estimated the business to be 100 million dollars in 1924. Vinal's third edition in 1940 and his "Primary Batteries" in 1950 brought the literature through World War II. G. J. Young added his "Fuel Cells" book in 1960, followed by the American Chemical Society (ACS) "Fuel Cell Systems," a report of a fuel cell symposium chaired by Young and Linden in 1965. This followed a fuel-cell volume of 1963 by the American Institute of Chemical Engineers.

"Regenerative Systems," another ACS volume, a report of a symposium of industrial and engineering chemists and Fuel Chemistry Divisions in 1965, was published in 1967.

Meanwhile, over decades the Bureau of Standards has been testing batteries, the American Society of Testing and Materials (ASTM) and American National Standards Institute (ANSI, formerly American Standards Association—ASA—and United States of America Standards Institute—USASI) have been developing and simplifying construction and standards, the National Electronic Distributors Association has been developing numbering systems and coordinating manufacturers' designation, and the Association of American Battery Manufacturers (AABM) has done a similar classification for secondary units. There are still, however, very large numbers of dry cells for myriad uses. The activity of the International Electrochemical Commission, embracing primarily the European countries, indicates that every European car maker feels it necessary to design a special hardware battery.

Meanwhile, in the United States the old battery makers grow and consolidate, new specialists appear, and electronic-components makers go into the battery business. The workhorses are still the dry cells of the primary type and the lead–sulfuric acid cells of the secondary type in tonnage, volume, and financial value; they are followed by the alkaline and mercury primaries, and the nickel-cadmium secondaries.

Military desires, which disregard economic considerations, have caused tremendously costly research programs, as have the air space race and international communication satellite projects.

chapter 2

The Voltage Concept—Standard Cells

Preparation

The intensity, difference of potential, or electromotive force (emf) is of particular interest in selection and arrangement of electrodes in batteries.

If the free energy (which might be roughly considered as the heat of reaction) be corrected for the temperature effect above the absolute zero (0° Kelvin) of a reaction per equivalent, then converted to electrical units (watts or volt-coulombs) and divided by the value of the faraday (coulombs), the quotient will be the first approximation of the emf in volts.

Properties

The reader might be more interested in application and engineering of battery systems than in the theory of development. Although outmoded, the Nernst concept of single electrode potentials might be useful in understanding battery systems, desirable and undesirable, and the magnitude of the voltage of such systems.

Nernst proposed a theory that the magnitude of a single electrode potential was a function of the "electrolytic solution pressure" of the metal and of the osmotic pressure of the metal ions in the solution. For equilibrium reactions involving anodic oxidation of the type in which a metal passes into solution and forms ions by the loss of electrons, or the reverse reaction at the cathode—i.e., the deposition of the metal and the gain of the electrons (valence increases represent loss of electrons, and valence decreases represent gain)—Nernst deduced the relation

$$n\mathrm{E}_s\mathfrak{F} = -RT \ln \frac{p}{P}$$

on the assumption that the gas laws are valid for ions of strong electrolytes.[1] For dilute solutions osmotic pressure varies with concentration. Then $p = kc$, where c in gram ions per liter is the concentration of metal ions and k is a constant. Then

$$n\mathrm{E}_s\mathfrak{F} = -RT \ln \frac{p}{kc}$$

$$\mathrm{E}_s = -\frac{RT}{n\mathfrak{F}} \ln \frac{p}{k} + \frac{RT}{n\mathfrak{F}} \ln c$$

For a given pure metal, the term $-(RT/n\mathfrak{F}) \ln (p/k)$ is a constant at a given temperature and may be written E_{so}, the normal electrode potential of the equilibrium in question. At 25°C, $T = 298°K$, $R = 8.32$ joules, and after converting natural to Briggs logarithms, the electrode potential for the cation is

$$\mathrm{E}_s = \mathrm{E}_{so} + \frac{0.059}{n} \log c$$

and for the anion

$$\mathrm{E}_s = \mathrm{E}_{so} - \frac{0.059}{n} \log c$$

In a normal solution c is unity and $\mathrm{E}_s = \mathrm{E}_{so}$. We then have a definition of electrode potential as the difference in potential between the electrode material and a normal solution (one gram ion per liter) of the ion in the equilibrium.

A table of electrode potentials such as Table 2 will furnish data in a concise form as to the quantitative aspect of electrode equilibriums.

When a cell operates reversibly the potential of the cell is represented by Eq. (1). For a reversible constant-temperature-and-pressure process, the work done by the system, not counting expansion work, is equal to $-\Delta F$ for the change in state. In a galvanic cell operating reversibly at

TABLE 2 Electrode Potentials of the Elements
(25°C, 1 N Ionic Concentration)

Reaction	E_0 against H^+ electrode	E_{abs}
$Cs^+ + e = Cs$	−3.02	−2.7374
$Li^+ + e = Li$	−2.957	−2.6744
$Rb^+ + e = Rb$	−2.924	−2.6414
$K^+ + e = K$	−2.922	−2.6394
$\frac{1}{2}(Sr^{++} + 2e = Sr)$	−2.92	−2.6374
$\frac{1}{2}(Ba^{++} + 2e = Ba)$	−2.90	−2.6174
$\frac{1}{2}(Ca^{++} + 2e = Ca)$	−2.87	−2.5874
$Na^+ + e = Na$	−2.712	−2.4294
$\frac{1}{2}(Mg^{++} + 2e = Mg)$	−2.40	−2.1174
$\frac{1}{3}(La^{3+} + 3e = La)$	−2.37	−2.0874
$\frac{1}{2}(Ti^{++} + 2e = Ti)$	−1.75	−1.4674
$\frac{1}{3}(Al^{3+} + 3e = Al)$	−1.7	−1.4174
$\frac{1}{2}(Be^{++} + 2e = Be)$	−1.69	−1.4074
$\frac{1}{4}(U^{4+} + 4e = U)$	−1.4	−1.1174
$\frac{1}{2}(Mn^{++} + 2e = Mn)$	−1.12	−0.8374
$\frac{1}{2}(Te + 2e = Te^-)$	−0.827	−0.5444
$\frac{1}{2}(Zn^{++} + 2e = Zn)$	−0.758	−0.4754
$\frac{1}{2}(Cr^{++} + 2e = Cr)$	−0.6	−0.3174
$\frac{1}{2}(S^- + 2e = S)$	−0.51	−0.2274
$\frac{1}{3}(Ga^{3+} + 3e = Ga)$	−0.5	−0.2174
$\frac{1}{2}(Fe^{++} + 2e = Fe)$	−0.44	−0.1574
$\frac{1}{2}(Cd^{++} + 2e = Cd)$	−0.397	−0.1144
$\frac{1}{3}(In^{3+} + 3e = In)$	−0.38	−0.0974
$Tl^+ + e = Tl$	−0.336	−0.0534
$\frac{1}{2}(Co^{++} + 2e = Co)$	−0.29	−0.0074
$\frac{1}{2}(Ni^{++} + 2e = Ni)$	−0.22	0.0626
$\frac{1}{2}(Sn^{++} + 2e = Sn)$	−0.13	0.1526
$\frac{1}{2}(Pb^{++} + 2e = Pb)$	−0.12	0.1626
$\frac{1}{2}(2D^+ + 2e = D_2)$	−0.0034	0.2792
$\frac{1}{2}(2H^+ + 2e = H_2)$	0.000	0.2826
$\frac{1}{3}(Sb^{3+} + 3e = Sb)$	0.10	0.3826
$\frac{1}{3}(Bi^{3+} + 3e = Bi)$	0.2	0.4826
$\frac{1}{3}(As^{3+} + 3e = As)$	0.30	0.5826
$\frac{1}{2}(Cu^{++} + 2e = Cu)$	0.344	0.6266
$Cu^+ + e = Cu$	0.51	0.7926
$\frac{1}{2}(I_2 + 2e = 2I^-)$	0.535	0.8176
$\frac{1}{3}(I_3 + 3e = 3I^-)$	0.54	0.8226
$\frac{1}{4}(Te^{4+} + 4e = Te)$	0.558	0.8406
$\frac{1}{2}(Hg_2^{++} + 2e = 2Hg)$	0.798	1.0806
$Ag^+ + e = Ag$	0.799	1.0816
$\frac{1}{2}(Pd^{++} + 2e = Pd)$	0.820	1.1026
$\frac{1}{2}(Hg^{++} + 2e = Hg)$	0.86	1.1426
$\frac{1}{4}(Pt^{4+} + 4e = Pt)$	0.863	1.1456
$\frac{1}{2}(Br_2 + e = Br^-)$	1.0648	1.3474
$\frac{1}{2}(Pt^{++} + 2e = Pt)$	∼1.2	1.4826
$\frac{1}{2}(Cl_2 + 2e = 2Cl^-)$	1.3583	1.6409
$\frac{1}{3}(Au^{3+} + 3e = Au)$	1.360	1.6426
$Au^+ + e = Au$	∼1.5	1.7826
$\frac{1}{2}(F_2 + 2e = 2F^-)$	1.90	2.1826

constant temperature and pressure, the change in free energy ΔF is

$$\Delta F = -nE\mathfrak{F}$$

where n = number of faradays involved, $\mathfrak{F}$ = faraday constant, E = electromotive force. Substitution gives

$$E = E° - \frac{RT}{n\mathfrak{F}} \ln \frac{(a_c)^c(a_D)^d}{(a_A)^a(A_B)^b} \qquad (1)$$

which expresses the variation of the emf of an electrochemical cell with the activity of the reactants and products of the cell and where E° represents the standard potential of the cell. If concentrations are substituted for the activities, the Nernst equation results. Most cells cannot be made to approximate reversible behavior. For others reversibility varies greatly. An electrode consists of an electronic conductor and the constituents of the surrounding electrolytic solution. All electrodes contain an element in two states of oxidation.

The relation between electrical energy of a system and the heat of reaction which has a temperature function is given by the Gibbs-Helmholtz equation

$$E + \frac{\Delta H}{n\mathfrak{F}} = T \frac{dE}{dT}$$

If E be expressed in volts and H in calories, then

$$E = \frac{-\Delta H}{n\mathfrak{F}} + T \frac{dE}{dT}$$

When dE/dT is positive, the emf of a reversible cell rises with temperature; when zero, the electrical energy is equal to the chemical energy.

Reference Electrodes

The emf of the cell is given by the difference in potential and is therefore equal to the difference of the single electrode potentials assuming zero electrical resistance of the components. Then

$$E = E_{s_1} - E_{s_2}$$

where E_{s_1} and E_{s_2} are single electrode potentials. If cells have their emf measured, the difference between any two single electrode potentials is determined. The absolute values are important in that they indicate whether a process occurring at a given electrode involves an increase or decrease of free energy. For convenience, the normal hydrogen electrode is chosen as the standard and assigned a zero single potential. The normal hydrogen electrode consists of platinized platinum immersed in an H_2SO_4 solution containing 1 g equivalent of hydrogen ion. Hydrogen is bubbled around this electrode and adsorbed or dissolved in the platinum to form a reproducible electrode. The emf of this electrode against the normal calomel (i.e., HgCl|Hg at 25°C) is -0.2822, and the absolute emf is then $+0.2826$. The emf of cells combining hydrogen electrodes with other electrodes are "single electrode potentials." The absolute potentials are therefore the potentials against the hydrogen electrode plus 0.2826. The potential E of an electrochemical reaction may be written

$$E = E_0 - \frac{0.05915}{N} \log_{10} \frac{(A_P)^x}{(A_R)^y}$$

where N is the number of faradays in the equation of the reaction, A_P and A_R the activities of the reactants, and x and y the corresponding coefficients in the electrochemical equation.

Any oxidation-reduction reaction may be broken up into two "half-reactions" or "couples." In the reaction

$$2Ag^+ + H_2 \rightleftharpoons 2Ag + 2H^+$$

the two half-reactions, or couples, are

$$Ag^+ + e \rightleftharpoons Ag \quad \text{and} \quad 2H^+ + 2e^- \rightleftharpoons H_2$$

Since any chemical reaction involves only the difference in potential between two couples, the absolute values are unnecessary. The hydrogen couple is an arbitrary zero reference for potentials of all other couples.

Reactions which involve hydroxide ion are also referred to as the hydrogen couple, but in such solutions this couple has the form

$$2H_2O + 2e^- \rightleftharpoons 2OH^- \qquad E_s = 0.828$$

The E_s values of half-reactions in alkaline solution will be designated as E_{sb} to indicate that this basic potential of the hydrogen couple must be used to obtain the completed reaction potential against hydrogen as

$$Cl^- + 2OH^- \rightleftharpoons ClO^- + H_2O + 2e^- \qquad E_{sb} = -0.94$$
$$ClO^- + H_2 \rightleftharpoons Cl^- + H_2O \qquad E_s = 0.83 + 0.94$$
$$= 1.77$$

Couples are written with the electrons on the left-hand side of the equation. A negative value for E_s will mean that the reduced form of the couple is a better reducing agent than H_2. For example

$$Zn^{++} + 2e^- \rightleftharpoons Zn \qquad E_s = -0.762$$

will mean that the reaction

$$Zn^{++} + H_2 \rightleftharpoons Zn + 2H^+$$

goes as written with a potential of -0.762 v. And similarly a positive E_s will mean that the oxidized form of the couple is a better oxidizing agent than H^+. For example

$$Cu^{++} + 2e^- \rightleftharpoons Cu \qquad E_s = +0.345$$

will mean that the reaction

$$Cu + 2H^+ \rightleftharpoons Cu^{++} + H_2$$

goes as written with a potential of 0.345 v.

The passage of ions from one valence to another is the oxidation potential. An ion of higher oxidation potential will oxidize one which shows a lower value. The ion of lower oxidation potential is the reducing agent. Thus, the same ion may serve as a reducing agent for one substance and an oxidant for others.

For a reaction of the type

$$M^y + (x - y)e \rightleftharpoons M^x$$

the potential difference varies with ionic concentration according to the equation

$$E_s = E_{so} + \frac{RT}{n\mathfrak{F}} \ln \frac{c}{c'}$$

where E_s is the emf produced by the reduction of M^x to M^y, c is the concentration of M^x, c' is the concentration of M^y, E_{so} is the emf of the cell when $c/c' = 1$, n is the difference in the valence, R the gas constant, T the absolute temperature, and $\mathfrak{F}$ the value of the faraday.

Table 3 gives representative values of oxidation-reduction potentials for some common systems.

In concentration cells, differences in potential exist between identical electrodes, one surrounded by a higher concentration of the same electrolyte than the other. If any liquid potential difference be neglected, $E = E_s - E_s'$, which becomes, when both electrodes give cations,

$$E = E_{so} - E_0' + \frac{0.058}{n} \log c - \frac{0.058}{n'} = \log c'$$

Since the electrodes and the ions are the same,

$$E = \frac{0.058}{n} \log \frac{c}{c'}$$

The emf is a function of the ratio of the ionic concentrations. These emfs seldom reach large values save when the concentration ratio is very great.

The emf of a concentration cell at ordinary temperatures, after correc-

TABLE 3 Standard Oxidation-Reduction Potentials

Reaction	E_0
$CNO^- + H_2O + 2e = CN^- + 2OH^-$	−0.97
$Fe(OH)_2 + 2e = Fe + 2OH^-$	−0.86
$H_2O + e = \tfrac{1}{2}H_2 + OH^-$	−0.828
$TlI + e = Tl + I^-$	−0.77
$HgS + H_2O + 2e = Hg + HS^- + OH^-$	−0.77
$HZnO_2^- + H_2O + 2e = Zn + 3OH^-$	−0.72
$Te + 2H^+ + 2e = H_2Te$	$\sim$−0.7
$Fe(OH)_3 + e = Fe(OH)_2 + OH^-$	−0.65
$PbO + H_2O + 2e = Pb + 2OH^-$	−0.58
$Se + 2H^+ + 2e = H_2Se$	$\sim$−0.5
$Ag(CN)_2^- + e = Ag + 2CN^-$	−0.5
$Cr^{3+} + e = Cr^{++}$	−0.4
$2H^+(10^{-7}M) + 2e = H_2$	−0.414
$Ti^{3+} + e = Ti^{++}$	−0.37
$Cu_2O + H_2O + 2e = 2Cu + 2OH^-$	−0.34
$PbSO_4 + 2e = Pb + SO_4^{--}$	−0.31
$H_3PO_4 + 5H^+ + 5e = P + 4H_2O$	−0.3
$Co(CN)_6^{--} + e = Co(CN)_6^{3-}$	−0.3
$V^{3+} + e = V^{++}$	−0.2
$CuI + e = Cu + I^-$	−0.17
$AgI + e = Ag + I^-$	−0.15
$2CuO + H_2O + 2e = Cu_2O + 2OH^-$	−0.15
$NO_3^- + 6H_2O + 8e = 9OH^- + NH_3$	−0.12
$Hg_2I_2 + 2e = 2Hg + 2I^-$	−0.04
$Ag_2S + 2H^+ + 2e = 2Ag + H_2S$	−0.036
$CuS + 2H^+ + 2e = Cu + H_2S$	−0.02
$NO_3^- + H_2O + 2e = 2OH^- + NO_2^-$	0.0
$HCNO + 2H^+ + 2e = HCN + H_2O$	0.0
$H_3SbO_3 + 3H^+ + 3e = Sb + 3H_2O$	$\sim$0.0
$WO_3 + 6H^+ + 6e = W + 3H_2O$	$\sim$0.0
$WO_3 + 4H^+ + e = WO^{3+} + 2H_2O$	$\sim$0.0
$TiO^{++} + 2H^+ + e = Ti^{3+} + H_2O$	0.04
$HgO + H_2O + 2e = Hg + 2OH^-$	0.099
$AgBr + e = Ag + Br^-$	0.10
$Hg_2Br_2 + 2e = 2Hg + 2Br^-$	0.13
$Sn^{4+} + 2e = Sn^{++}$	0.13
$SO_4^- + 4H^+ + 2e = H_2O + H_2SO_3$	0.14
$Cu^{++} + e = Cu^+$	0.17
$S + 2H^+ + 2e = H_2S$	0.17
$Ta_2O_5 + 10H^+ + 10e = 2Ta + 5H_2O$	$\sim$0.2
$PtCl_4^- + 2e = Pt + 4Cl^-$	$\sim$0.2
$AgCl + e = Ag + Cl^-$	0.223
$H_3AsO_3 + 3H^+ + 3e = As + 3H_2O$	0.24
$MoO_3 + 6H^+ + 6e = Mo + 3H_2O$	0.25
$Hg_2Cl_2 + 2e = 2Hg + 2Cl^-$	0.270
$PbO_2 + H_2O + 2e = PbO + 2OH^-$	0.3
$VO^{++} + 2H^+ + 4e = V + H_2O$	0.3
$VO^{++} + 2H^+ + e = V^{3+} + H_2O$	0.4
$O_2 + 2H_2O + 4e = 4OH^-$	0.40
$PtCl_6^{--} + 2e = PtCl_4^{--} + 2Cl^-$	$\sim$0.40
$UO_2^{++} + 4H^+ + 2e = U^{4+} + 2H_2O$	0.41
$H_2SO_3 + 4H^+ + 4e = S + 3H_2O$	0.47
$Fe(CN)_6^{--} + e = Fe(CN)_6^{3-}$	0.49

TABLE 3   Standard Oxidation-Reduction Potentials (Continued)

Reaction	E_0
$H_3AsO_4 + 2H^+ + 2e = H_3AsO_3 + H_2O$	0.49
$NiO_2 \cdot 2H_2O + 2e = Ni(OH)_2 + 2OH^-$	0.49
$Ag_2CO_3 + 2e = 2Ag + CO_3^{--}$	0.50
$MoO_3 + 4H^+ + e = MoO^{3+} + 2H_2O$	0.5
$I_2 + 2e = 2I^-$	0.535
$I_3 + 2e = 3I^-$	0.54
$2HgCl_2 + 2e = Hg_2Cl_2 + 2Cl^-$	0.63
$MnO_4^- + e = MnO_4^{--}$	0.66
$O_2 + 2H^+ + 2e = H_2O_2$	0.68
$AgBrO_3 + e = Ag + BrO_3^-$	0.68
$C_6H_4O_2$ (quinone) $+ 2H^+ + 2e = C_6H_4(OH)_2$	0.70
$MnO_4^- + 2H_2O + 2e = MnO_2 + 4OH^-$	0.71
$H_2SeO_3 + 4H^+ + 4e = Se + 3H_2O$	0.74
$H_3SbO_4 + 2H^+ + 2e = H_3SbO_3 + H_2O$	0.75
$Cu^{++} + I^- + e = CuI$	0.85
$HNO_2 + 7H^+ + 6e = 2H_2O + NH_4^+$	0.86
$HO_2^- + H_2O + 2e = 3OH^-$	0.87
$CoO_2 + H_2O + 2e = CoO + 2OH^+$	0.9
$ClO^+ + H_2O + 2e = Cl^- + 2OH^-$	0.94
$NO_3^- + 4H^+ + 3e = NO + 2H_2O$	0.94
$NO_3^- + 3H^+ + 2e = HNO_2 + H_2O$	0.95
$HNO_2 + H^+ + e = NO + H_2O$	0.98
$HIO + H^+ + 2e = I^- + H_2O$	0.99
$OsO_4 + 4H^+ + 4Cl^- + 2e = OsO_2Cl_4^{--} + 2H_2O$	$\sim$1.0
$IO_3^- + 6H^+ + 6e = I^- + 3H_2O$	1.09
$HVO_3 + 3H^+ + e = VO^{++} + 2H_2O$	1.1
$Tl^{3+} + 2e = Tl^+$	1.2
$H_2SeO_4 + 2H^+ + 2e = H_2SeO_3 + H_2O$	$\sim$1.2
$O_2 + 4H^+ + 4e = 2H_2O$	1.23
$PdCl_6^{--} + 2e = PdCl_4^{--} + 2Cl^-$	1.3
$HCrO_4^- + 7H^+ + 3e = Cr^{3+} + 4H_2O$	1.3
$HBrO + H^+ + 2e = Br^- + H_2O$	1.33
$MnO_2 + 4H^+ + 2e = Mn^{++} + 2H_2O$	1.33
$ClO_4^- + 8H^+ + 8e = Cl^- + 4H_2O$	1.35
$Au_2O_3 + 6H^+ + 6e = 2Au + 3H_2O$	1.362
$IO_4^- + 8H^+ + 8e = I^- + 4H_2O$	1.4
$BrO_3^- + 6H^+ + 6e = Br^- + 3H_2O$	1.42
$PbO_2 + 4H^+ + 2e = Pb^{++} + 2H_2O$	1.44
$ClO_3^- + 6H^+ + 6e = Cl^- + 3H_2O$	1.45
$HClO + H^+ + 2e = Cl^- + H_2O$	1.50
$H_2S_2O_8 + 2e = 2SO_4^{--} + 2H^+$	$\sim$1.5
$CeO_2 + 4H^+ + e = Ce^{3+} + 2H_2O$	1.5
$MnO_4^- + 8H^+ + 5e = Mn^{++} + 4H_2O$	1.52
$MnO_4^- + 4H^+ + 3e = MnO_2 + 2H_2O$	1.63
$FeO_4^{--} + 8H^+ + 3e = Fe^{3+} + 4H_2O$	$\sim$1.7
$HBiO_3 + 5H^+ + 2e = Bi^{3+} + 6H_2O$	$\sim$1.7
$PbO_2 + 4H^+ + SO_4^{--} + 2e = PbSO_4 + 2H_2O$	1.7
$H_2O_2 + 2H^+ + 2e = 2H_2O$	1.78
$NiO_2 \cdot 2H_2O + 4H^+ + 2e = Ni^{++} + 4H_2O$	1.8
$O_3 + 2H^+ + 2e = O_2 + H_2O$	1.9

tion for liquid junction potential, shows a deviation from the value calculated from the formula where the values of ionic concentration are obtained from conductivity data. Strictly, the activity a and a' of the ions should be substituted. At extreme dilution the values are identical, but they differ to some extent at moderate concentrations.

Single electrode potentials are measured by combining an electrode with either the standard hydrogen electrode or some other whose potential is known on the hydrogen scale. The emf is then potentiometric.

The standard electrodes are the normal and 0.1 normal calomel electrodes in which the system is Hg|KCl solution saturated with HgCl. Such a half cell is shown in Fig. 3. The potential of the normal calomel electrode Hg|HgCl N KCl is $+0.286$ v at 18°C and that of the decinormal, Hg|HgCl 0.1 N KCl, is $+0.338$ v at 18°C. For alkaline solutions the cells Hg|HgO N NaOH (E = $+0.117$ v at 18°) and Hg|HgO 0.1 N NaOH (E = $+0.72$ v at 18°) are useful.

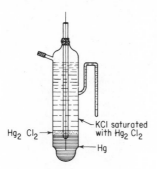

Fig. 3 Half cell.

With acid solutions, the standard hydrogen electrode (E = ± 0.0 v) is satisfactory, but the two systems Hg|Hg$_2$SO$_4$ N H$_2$SO$_4$ (E = $+0.685$ v at 18°) and Hg|Hg$_2$SO$_4$ 0.1 N H$_2$SO$_4$ (E = $+0.687$ v at 18°) are more convenient.

Connection between the electrode being measured and the standard electrode is made by a bridge of a high-conductivity electrolyte such as KCl in the case of the normal calomel electrode.

It is sometimes desirable to measure the difference in potential at an electrode during the course of electrolysis. The siphon side tube of the half cell is lengthened, bent horizontally at the end, and drawn to a fine point which can be brought up against the electrode. The error of the potential drop caused by the current is low, seldom exceeding 0.002 v.

"Standard" and/or reference cells are electrochemical systems which serve as a standard of emf. With standards of resistance, the standard cell is also employed for the measurement of current. Standard cells are not "standardized" in respect to form, shape, dimensions, application, or output factors which will be discussed in connection with batteries or electrochemical power sources. Standard cells are not power sources and are never employed for the delivery of current. Were they so employed, their usefulness would be destroyed, so that in measurements work they are applied only by "null methods," balanced resistances, Wheatstone bridges, and the like. They are not batteries or power sources.

When measurements of electric power are made in terms of standards of emf and resistance, the expression for power, $P = E^2/R$, emphasizes knowing E accurately, since an error in the standard for E would produce a percentage error twice as great in the value for the power P.

The emf of each standard cell is certified by its manufacturer as of a certain date, and many cells are certified also by the National Bureau of Standards (NBS) when direct comparisons are made with the national standard. The unit of emf is known as the volt, or as the absolute volt, to distinguish it from the international volt in use prior to 1948.

The absolute system of electrical units is derived from fundamental mechanical units of length, mass, and time by principles of electromagnetism, with the permeability of space taken as unity in the centimeter-gram-second (cgs) units or as 10^{-7} in the corresponding meter-kilogram-second (mks) units. Electrical measurements are thus made concordant with measurements in other fields of science and engineering.

Electrical energy is a function of two factors: the quantity or current (amperes), and the intensity or difference in potential (volts). The extent of a change occurring in any energy content of a system is determined by both factors, but the possibility of a change is determined only by the difference in potential. Hydraulically, water can move from a higher to a lower level, depending upon the relative heights of the two levels. The potential energy change in the system is the product of the weight which has passed from one level to the other and the difference in height. Similarly, temperature difference is a thermal potential which will determine the heat transference, while affinity is a chemical potential. The product of the potential factor, or affinity of the reaction, and the quantity factor, or amount of matter which has been transformed, gives the quantity of energy involved in the transference.

The unit of electrical current is the ampere.[2] When electricity is passed through an aqueous solution of a metallic salt, the salt is decomposed. In many cases the metal is deposited in the free state. The ampere was defined until 1948 as the unvarying electric current which will deposit silver at the rate of 0.00111800 g/sec from a solution of $AgNO_3$ in water under a given set of conditions.[3] The current density per unit of cross-sectional area is the current flowing through a conductor divided by the area of the conductor. Units are amperes per square foot (asf) or amperes per square decimeter (asd).

The standardizing agencies of the world for half a century tried to develop an absolute scale of electrical energy units based on mechanical units. This was achieved in 1948. The ampere is now defined in absolute units as an electric current of such magnitude that when maintained in two straight parallel conductors of infinite length and of negligible cross section, at a distance of one meter from each other in vacuum, it would produce between the conductors a force of 2×10^{-7} newton per meter of their length.

A newton is the mks unit of force that will give a mass of one kilogram (kg) an acceleration of one meter per second per second (1 m/sec^2), and is equal to 10^5 dynes.

The mks system of units is related to the cgs system. The magnitudes of many derived units are more convenient for engineering purposes.

The dyne is the cgs unit of force which will give a mass of one gram (1 g) an acceleration of one centimeter per second per second (1 cm/sec^2). The erg is the cgs unit and is the work done when a force of one dyne acts for a distance of one centimeter. The joule is the mks unit of work done when a force of one newton acts for a distance of one meter (1 m). The joule is equal to 10^7 ergs. The watt is the mks unit of power required to do work at the rate of one joule per second (1 joule/sec) or 10^7 ergs per second (10^7 ergs/sec).

Electric current (I) is defined as the rate at which electricity flows through a conductor or circuit, in amperes, which is a current of one coulomb per second (1 coul/sec). The cgs unit is either the abampere or statampere. Electric current density is the ratio of the current flowing through a conductor to the cross-sectional area of that conductor. The cgs unit is either the abampere or statampere per square centimeter.

Electric quantity (Q) is defined as the amount of electricity present in any electric charge or passed through a circuit during any time interval by an electric current. The unit is the coulomb; the cgs units are the abcoulomb and statcoulomb.

The quantity of current is a function of both the current strength and time. The unit is the coulomb[4] or ampere-second, defined until 1948 either as the quantity of electricity passing in one second at a current strength of one ampere, or, from the definition of the ampere, being that quantity of electricity which, when passed through a solution of $AgNO_3$, will deposit 0.00111800 g of silver. Now the coulomb is the quantity of electricity transported in one second across any cross section of a circuit by a current of one ampere.

Electrical resistance (R) is the quantity in ohms, analogous to friction, in a conductor that measures the difference in potential required to maintain a given electric current through it. The cgs unit is either the abohm or statohm.

Electric resistivity (ρ) is the ratio of potential gradient in a conductor to the current density thereby produced as well as the specific resistance of a substance, numerically equal to the resistance offered by a unit cube of the substance as measured between a pair of opposed parallel faces, the unit being the ohm-centimeter (ohm-cm). The cgs electromagnetic unit is the abohm-centimeter.

Electric conductance (G) is the conducting power of a conductor or circuit for electricity or the inverse or reciprocal of electrical resistance. The unit is the mho. The cgs unit is either the abmho or statmho.

Electric conductivity (γ) is the specific electric conducting power of a substance or the reciprocal of resistivity in mhos per centimeter (mho/cm). The cgs unit is the abmho per centimeter.

The ohm[5] was defined until 1948 as equal to 1,000 million units of resistance of the cgs system of electromagnetic units and represented by the resistance offered to an unvarying electric current by a column of mercury at the temperature of melting ice, 14.4521 grams in mass, of a constant cross-sectional area, and of the length of 106.3 centimeters. In effect, this means a column of mercury 1 mm^2 in cross section, 106.3 cm long. The ohm is now defined as the electric resistance between two points of a conductor when a constant difference of potential of one volt maintained between these points produces a current of one ampere in the conductor, the conductor not being the source of any emf.

The intensity factor "electromotive force" or "difference in potential" is the volt.[6] Electromotive force (emf) or pressure is that which tends to make an electric current flow. Difference in potential causes a tendency of an electric current to flow from a level of higher to one of lower potential. The numerical measure of the potential difference is the work done on a unit quantity of electricity in passing between the two points.

Potential gradient is the space rate of change of potential, or the rate of change with respect to distance, the unit being volt per meter (v/m). The cgs unit is either the abvolt or statvolt per centimeter.

The volt before 1948 was "the electrical potential difference which, when steadily applied to a conductor having a resistance of one ohm, will produce in it a current of one ampere."[7] Now the volt is the difference of electrical potential between two points (or two equipotential surfaces) in a conductor carrying a constant current of one ampere when the power dissipated between these points is one watt. The volt cannot be easily produced as defined, owing to the definition of the ampere. The emf of a voltaic cell, however, can be determined against the international ohm and the international ampere, and such a cell can be used as a medium for realizing the international volt. "Standard cells" have provided the means for making comparisons of emf, giving a definition of volt as the emf which, steadily applied to a conductor whose resistance is one international ohm, will produce a current of an international ampere and is practically equivalent to 1,000/1,434 of the emf between the poles or electrodes of the Clark cell at 15°C.

Without detailing the tremendous amount of work of international agencies, societies, and standardizing committees, the relation may be seen in Fig. 4. The left half represents functions of NBS, the right half the units and certified standards used by the public.

A considerable number of cells have been proposed as standards of emf, and some of them have appeared in a variety of forms. The most important of these are listed on Table 4. Only the Clark cell of 1872 as modified by Lord Rayleigh and the Weston cell of 1893 with its later modifications were of importance.

Weston's cadmium cell of 1893 in its two forms, with saturated and unsaturated cadmium sulfate electrolyte, has superseded all earlier forms of standard cells. The unsaturated form is commercially known as a "standard cell." The saturated cell is made of the same basic materials and is capable of higher precision. For this reason it is preferred by the national standardizing laboratories and other institutions which have the facilities it requires. Unsaturated cells are more portable and convenient. They do not require thermostatic control and they are sufficiently accurate for all but the most precise measurements. Many thousands of the unsaturated cells are in daily use.

In the Eppley Laboratory form the container of the saturated cell is an H-shaped glass vessel with platinum wires sealed in at the lower ends of

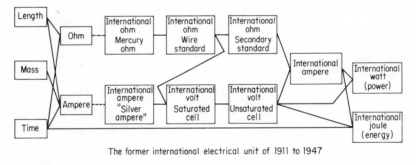

The former international electrical unit of 1911 to 1947

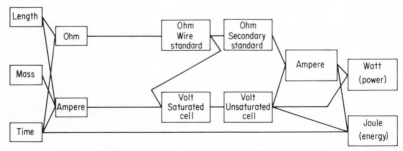

The present absolute electrical units in use since January 1, 1948

Fig. 4 Diagram showing the relation of the present and former systems of electrical units to the basic mechanical units of length, mass, and time.

TABLE 4 Various Types of Standard Cells

Name	Date	System	Approximate emf, volts
Daniell	1836	Zn|ZnSO$_4$|CuSO$_4$|Cu	1.079
Clark	1872	Zn|ZnSO$_4$|Hg$_2$SO$_4$|Hg	1.433 at 15°C
De la Rue	1878	Zn|ZnCl$_2$|AgCl|Ag	1.03
Helmholtz	1882	Zn|ZnCl$_2$|HgCl|Hg	1.000 at 15°C
Weston-Clark	1884	Zn|ZnSO$_4$|Hg$_2$SO$_4$|Hg*	(No record)
Gouy	1888	Zn|ZnSO$_4$|HgO|Hg	1.390 at 12°C
Carhart-Clark	1889	Zn|ZnSO$_4$|Hg$_2$SO$_4$|Hg	1.44 at 15°C
Weston	1893	Cd|CdSO$_4$|Hg$_2$SO$_4$|Hg†	1.0183 at 20°C

* Unsaturated electrolyte.
† Also an unsaturated form.

the vertical legs for electrical connection. In one leg there is mercury and in the other, cadmium amalgam. The mercury electrode is covered with a mixture of mercurous sulfate and finely ground cadmium sulfate crystals. A layer of larger cadmium sulfate crystals is placed upon the surface of both the amalgam and the mercurous sulfate. The cell is filled to above its crossarm with a saturated solution of cadmium sulfate after which the open ends of the H-tube are sealed by fusing in a flame to ensure airtightness. Mercurous sulfate is slightly light-sensitive and exposure of cells to light should be limited to short, infrequent periods.

The Eppley Laboratory produces both the nonshippable and shippable types of saturated cells. An H-shaped cell is $4\frac{3}{4} \times 2\frac{1}{2} \times 5$ in. for establishing a primary reference volt, or a miniature, $3\frac{3}{4} \times 1 \times \frac{3}{8}$ in. These two types must be hand-carried when transported, as damage will occur to the cells if they are not maintained in an upright position.

A miniature shippable cell provides a reliable voltage reference accurate to 0.001%, which combines the high stability and long life of a saturated cell with the portability that is a feature of unsaturated cells. Shippability is achieved through a septum which retains the electrode materials in place.

For a high-accuracy secondary standard, the unsaturated cadmium standard cell has been accepted. This is similar in form to the normal cell except that the cadmium sulfate solution is unsaturated at room temperatures, no excess of the solid cadmium sulfate being added to either the mercurous sulfate or the solution. The container of the unsaturated cell is also an H-shaped glass vessel.

The temperature coefficient of a cell is the algebraic sum of the temperature coefficients of the two electrodes, the emf of the mercurous sulfate or positive limb being positive, and that of the cadmium amalgam limb being negative. Therefore, it is important that both electrodes be held at the same temperature.

The principal advantage of the unsaturated cell is its small net temperature coefficient which averages only $-3\ \mu v/°C$ between 4 and 40°C (see Fig. 5).

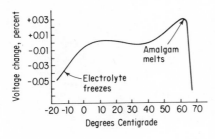

Fig. 5 Temperature effect of standard cells. (*The Eppley Laboratory, Inc.*)

When maintained at normal laboratory temperatures and operative conditions, the usable life of unsaturated standard cells is from 5 to 10 years.

All Muirhead cells utilize an acid–cadmium sulfate electrolyte, with cadmium-mercury amalgam in the negative limb, and mercury (together with a layer of mercurous sulfate) in the positive limb. The electrolyte is saturated with cadmium sulfate and in addition, contains an excess of solid cadmium sulfate crystals; these ensure that the electrolyte remains saturated at all temperatures. Reference cells, however, contain an electrolyte consisting of a solution of cadmium sulfate that is incompletely saturated at temperatures exceeding 4°C.

Construction follows the H pattern, variations in the design producing maximum mechanical strength together with minimum internal resistance and freedom from air locks. The solid chemical constituents are secured in position by chemically inert porous septa of sintered, high-molecular-weight polythene. Figure 6 shows an unmounted precision cell, Fig. 7 a mounted unit, Fig. 8 an unmounted reference cell, and Fig. 9 an unmounted miniature cell.

The miniature standard cell has small size and low cost which are advantages over the zener diode. With solid-state circuitry the input impedance of many applications is normally in excess of 10,000 megohms, which means that the current withdrawn from the cell is of a very low order. The small dimensions of the cell make it suitable for mounting directly onto printed circuit boards or in an adjacent position of minimum temperature variation.

The miniature is a single tube, dipcoated in red PVC for protection. It is fitted with insulated flying leads 3 in. long and colored red and black

The Voltage Concept—Standard Cells   23

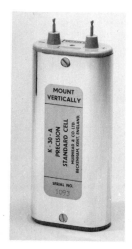

Fig. 6 An unmounted precision standard cell. (*Muirhead Instruments, Inc.*)

Fig. 7 A mounted precision standard cell. (*Muirhead Instruments, Inc.*)

to denote polarity. Each cell is marked with month and year of manufacture.

The Weston cell's form, of which Weston Instruments, Incorporated, was the original producer under the Weston patents now expired, has become stabilized with several manufacturers. The references detail the

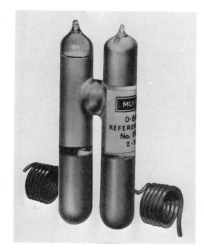

Fig. 8 An unmounted reference cell. (*Muirhead Instruments, Inc.*)

Fig. 9 An unmounted miniature standard cell. (*Muirhead Instruments, Inc.*)

history.[8] Work on standard cells was done by Weston Instruments in the areas of cell drift, thermal hysteresis of cell emf, and cell resistance to vibration and shock. Cell appearance remained the same.

Other developments were concerned with miniaturization of cells for industrial purposes and with efforts to provide cells in configurations of the coaxial type, in which one leg of the cell surrounds the other.

Bradd and Eiche[9] published a paper in 1962 detailing their findings after unsaturated cells were subjected to a series of shock and vibration tests. The effect of mechanical vibration and shock on a group of four 500-ohm unsaturated standard cells was investigated (two each from separate manufacturers). Vibration studies were conducted at frequencies of 10, 20, 30, 50, 70, 100, 200, 500, and 1,000 Hertz (Hz) with accelerations of 1, 2, 5, 7, and 10 g. The shock studies were conducted with shock durations of 6.2, 11, and 18 milliseconds (msec) and shock accelerations of 10, 20, 30, and 40 g. Transient ac and dc cell outputs were observed related to the magnitude of the acceleration and the frequency of the vibration. The observed shift in dc emf in the cells was within the 0.01% limit of precision usually assigned to cells of this type, for all frequencies at 1 g acceleration. For 10 g acceleration, the ac and dc transient effects exceeded the 0.01% level. The ac output of the cell was large and exceeded 0.1 mv on many occasions. The shocks were almost without effect on the dc emf. The cells were within the 0.01% level of precision during the shock studies: indeed, the shift in dc emf was less than 2 μv. However, test cell output transients during the shock study often exceeded 1 mv. After shock or vibration, the cells recovered within the 0.01% level of precision, and returned to the preshock or vibration condition in less than 3 min. The unsaturated standard cell did not behave as a "delicate" standard.

According to Weston Instruments, the fully saturated standard cell has traditionally been the most stable voltage reference available when maintained in a closely controlled environment and carefully handled to prevent inversion or shock and vibration of any magnitude. The "portable saturated cell with septum" appears to have overcome some of these latter difficulties.

The model 4, type 3 unsaturated form of cadmium cell has been redesigned to improve its characteristics. These are:

1. Electromotive force—between 1.01880 and 1.01950 absolute volts. (Cells made after July 1965 have values between 1.01900 and 1.91940 absolute volts.)
2. Cell resistance—typically between 70 and 90 ohms.
3. Temperature coefficient—typically between $+5$ ppm/°C and -5 ppm/°C.

4. Hysteresis—recovery typically within 6 ppm (0.0006%) within 4 hr when subjected to a sudden temperature drop of 7°C (12.6°F).

5. Long-term stability—typically 24 μv decrease in value per 6-month interval.

6. Cell certified to 0.05%.

REFERENCES

1. Where p is the electrolytic solution pressure of metal M, P the osmotic pressure of the metal ions in solution, E_s the single electrode potential corresponding to the equilibrium, R the gas constant, and T the absolute temperature, $\mathfrak{F}$ the value of the faraday in coulombs.
2. Named after A. M. Ampère (1775–1836), French physicist and chemist.
3. It is specified that the $AgNO_3$ solution shall contain 15 to 20 g of salt to 100 g of distilled water. The solution must be used only once, not less than 100 cc at a time, and not more than 30% of the metal must be deposited. The current density must not exceed 0.02 asc at the cathode, and 0.2 asc at the anode.
4. Named after C. A. Coulomb (1736–1806), French physicist.
5. Named after G. S. Ohm (1787–1854), German physicist.
6. Named after Count A. Volta (1745–1827), Italian physicist.
7. "International Critical Tables," vol. 1, p. 34, McGraw-Hill Book Company, New York, 1926.
8. a. W. J. Hamer, History of the National Standard of Electromotive Force, *NBS* (20 references) *ISA Paper No.* M 2-1 *MESTIND*-67, 22d Annual *ISA* Conference Proceedings, vol. 22, part 1, 1967.

    b. A. R. Karoli, Capabilities of Modern Unsaturated Cells (6 references), *ISA* Paper No. M 2-3 *MESTIND*-67.

    c. Standard Cell, Modified Weston, Over Long Periods, *NBS Research Paper* No. 5971; Standard Cells, Saturated, Oil Baths, *NBS Research Paper* 4799; Standard Cells, Their Construction, Maintenance and Characteristics, Mono. 84, *NBS Mono.* (*J. of Research, NBS*).
9. Bradd and Eiche, Effect of Vibration and Shock on Unsaturated Cells, *J. Res. Nat. Bur. Std., Engineering and Instrumentation*, 66C, no. 2, (April–June 1962).

chapter 3
The Current-producing Cells and Batteries

Chemelectrics concerns itself with the conversion of the chemical energy into electric current. It was less important than the inverse operation in which halogens or metals are separated by or refined electrolytically. However, transistorized and miniaturized equipment have such low power requirements that batteries have become competitive with other sources. The requirements of airborne and underwater instruments, missiles, and controls, have spurred the development of many different battery types.

Comparative and General

Fuel cells have been developed to convert the chemical energy obtained by the combustion of fuels directly into electrical energy for military use where economics are secondary. Transformations of solar radiation and of the energy of radioactive isotopes into electrical forms are also related to this development.

The conversion of chemical into electrical energy is a function of cells or batteries, subdivided into those designated as primary, or nonregenerated for reuse, and secondary, readily reversible and rechargeable. This distinction is important primarily to industry, as the Daniell or gravity and the Lalande caustic early primary cells, are reversible to a high degree, but these have disadvantages preventing their use as storage batteries.

A cell designates a single unit but is applied to an assembly of identical units. A battery is strictly an arrangement of two or more cells, usually connected in series or parallel to supply the necessary current or voltage or both. Batteries are for the intermittent production of small amounts of electrical energy, as for electrotherapeutic purposes, electric bells, signal systems, lighthouses, small motors used intermittently, control apparatuses, telephone systems, motor ignitions, portable radio sending and receiving sets, and self-winding watches. More efficient and larger forms can be employed for isolated installations of electric lighting and the driving of small motors. The high cost of primary cells and batteries makes production of electricity in large quantities impractical. One of the most efficient units, the air cell, is estimated to produce 1 kwh during its useful life at a cost of \$5/kwh.

Before and during World War II there developed an urgent need for special-purpose batteries of high capacity (with the disadvantage of short life), high voltage, light weight, and other characteristics for radiosonde, guided missiles, pilot balloon lights, electronic controls, control radio operation, and the like, as well as short-time operation of special motors and other mechanisms. Capacity is the output capability over a period of time, in ampere hours (ah). Large quantities of magnesium–silver chloride, magnesium–copper chloride cells were manufactured in a wide range of sizes, capacities, and voltages.

In practice, electrode systems never behave absolutely reversibly. Passivity is avoided by soluble anodes and controlled electrolytes, with anode corroding agents. "Anode" and "cathode" are employed here with reference to the inside of the cell, while externally the cathode becames the positive plate and the anode the negative one. When current is drawn from the cell and the unit serves as an energy source, the positive terminal is the projection of the cathode, and the negative terminal is the projection of the anode. At the cathode in simple cells there is a reversible discharge of metallic ion to metal. In others, such an indifferent electrode surrounded by a depolarizing, oxidizing agent may constitute the cathode at which H^+ ions are discharged. The depolarizers increase the emf of a cell above the normal reversible H^+ ion discharge. The electrode does not become saturated with hydrogen owing to the speed of the reaction between the depolarizer and the hydrogen. The units are operated to a cutoff voltage, the value at the end of useful discharge.

A depolarizer must be a good conductor and make effective contact with the electrode as well as react rapidly with the hydrogen. In older

TABLE 5 Commercially Important Battery Systems

Type	System	Anode	Cathode	Electrolyte	Internal resistance, ohms	Operating temperature, °F	Output voltage, open-circuit	Maximum current drain, ah	Capacity, ah	wh/lb	wh/cu in.	Cost/wh, cents	Cost/kwh generated, dollars
Dry cell	$Zn\|NH_4Cl\|ZnCl_2\|MnO_2\|C$	Zn	$MnO_2\|C$	$NH_4Cl\|ZnCl_2$	0.3–0.5	0–100	1.5	0.03–2	5	26	1.6–2.0	3	30.00
Alkaline	$Zn\|KOH\|MnO_2\|Fe$	Zn	$MnO_2\|C$	KOH	0.05–0.2	−40–120	1.5	1.3–30 (sc)	20	30	3.5	5.2	52.00
R.M. mercury	$Zn\|KOH\|HgO\|Hg$ or $Zn\|KOH\|HgOMnO_2$	Zn	$HgO\|Fe$	KOH	0.25–10	40–165	1.35–1.4	1–10 (sc)	14	35	3.3	9.8	96.00
Silver	$Zn\|KOH\|Ag_2O\|Fe$	Zn	$Ag_2O\|Fe$	NaOH / KOH	2.7–5.5	−20–100	1.6	80 ma	0–0.10	75	6	24	240.00
Air cell	$Zn\|NaOH\|O_2\|C$	Zn	$O_2\|C$	NaOH			1.4–1.6	0.1–2	1000	80–100	3–5	0.9–1.0	9.00–10.00
Mg-AgCl	$Mg\|AgCl\|Ag$	Mg	$AgCl\|Ag$	$MgCl_2$	0.1–1.0	−50–65	1.4–1.5	10–13	20	19–20	8.54	9	90.00
Mg-CuCl$_2$	$Mg\|CuCl_2\|Cu$	Mg	$CuCl_2\|Cu$	$MgCl_2$	0.2–1.2	−50–65	1.4–1.6	10–13	20	18–20	8.5	8	80.00
Mg dry cell	$Mg\ alloy\|MgBr_2\|MnO_2\|C$	Mg	$MnO_2\|C$	$MgBr_2$			1.8						150.00

types of cells liquid depolarizers were used. They had the advantage of rapid reaction, but the disadvantage of diffusion toward, and attack on, the anode. Diaphragms were often necessary and the cell resistance was increased. The more effective the depolarizers, the more closely will the cell system approach the equilibrium or theoretical value.

Table 5 lists the commercially important battery systems and compares their systems, voltages, weights per unit of capacity, and applications. Table 6 lists potential combinations of anode and cathode materials.

Low or even freezing temperatures are not harmful to zinc-carbon cells as long as there is not repeated cycling from low to higher temperatures. Low-temperature storage is beneficial to shelf life. A storage temperature of 40 to 50°F is effective. When batteries are removed from low-temperature storage, they should be allowed to reach room temperature in their original packing to avoid condensation of moisture which may cause electrical leakage and destruction of the jackets.

There is no relation between continuous-duty service and intermittent service. It is, therefore, impossible to rate the merits of different systems on intermittent service by comparing results of continuous-duty tests.

The short-circuit amperage of a cell may be adjusted by varying the carbon content of the depolarizing mix. Carbon contributes nothing to the cell energy but reduces resistance.

To the electrometallurgist electrowinning metals, the zinc–manganese dioxide cell represents the back emf of a cell from which he is trying to obtain zinc, and he is subject to power interruption or failure. The entire "line" is a huge primary battery with a sulfuric acid electrolyte. In a similar fashion in a manganese metal plant, power interruption gives

TABLE 6 Comparison of Various Possible Electrode Materials

	Anode materials				Cathode materials				
Metal	emf sl. acid, v	emf alkaline, v	mah/g	ah/cc	Material	emf sl. acid, v	emf alkaline, v	mah/g	ah/cc
Magnesium	−2.36	−2.69	2,204	3.84	MnO_2	+0.80	+0.27	307	1.54
Aluminum	−1.67	−2.35	2,982	8.05	PbO_2	+1.46	+0.25	240	2.26
Zinc	−0.76	−1.25	820	5.65	HgO	+0.85	+0.10	248	2.76
Iron	−0.44	−0.88	960	7.55	CuO	+0.34	−0.36	670	4.32
Cadmium	−0.40	−0.81	477	4.13	Ag_2O	+0.80	+0.35	232	1.67
Tin	−0.14	−0.91	452	3.30	AgO	ca. +1.9	+0.57	432	3.22
Lead	−0.13	−0.54	259	2.94	AgCl	+0.22	+0.22	187	1.04

him a huge manganese metal|alkaline manganese sulfate|acidic manganese sulfate–sulfuric acid|stainless steel battery, with destruction and disappearance of his product, electrolytic manganese.

Economics

Nearly every conceivable "system" has been worked on, developed, and studied. Only a few systems have been developed into "hardware," engineered, and mass produced.

For civilian application, batteries are an expensive source of power, justified by convenience, portability, self-containment, elimination of central power stations, connecting wires, and conductors, as well as ability to perform remotely without attention or inspection.

The average homeowner's charge for the power for his home full of electric appliances, lights, refrigerators, mixers, dishwashers, blenders, etc., is of the order of 1.2 ¢/kwh. He is willing to pay of the order of 100 times this amount for cordless appliances which he can recharge: for flashlights, controls, lighthouses, radios, signaling systems and the like in his daily life.

For military applications, economic considerations play no part. Power in a missile, a radiosonde, a communication satellite, or a space capsule might have no limitation as to cost per kwh. This can be several thousand to a million times the cost of power to the homeowner. This is emphasized in a paper of NASA's fuel-cell program.

Cohn[1] states that the goal of NASA's fuel-cell program is to obtain lightweight, dependable power sources for a variety of needs. These may include communication; command and control; guidance; radar; image acquisition, processing, and transmission; data handling and storage; life support; experiments on planetary surfaces and environment; and power for surface-exploration vehicles.

Among the major factors to be considered in designing space-type fuel cells are (1) the need for very high reliability, since chances for repair are extremely limited even on manned missions; (2) high energy and power densities, because it costs between $1,000 and $5,000 to put a pound of substance into space, and lift capabilities are limited while power requirements keep increasing; (3) the space environment (where gravity is absent) and the planet surface (which varies from that on earth), where radiation and meteoroids present hazards, where temperatures can fluctuate widely, and where there is no atmosphere providing oxygen to act as a heat sink.

The low-temperature fuel cell, with an ion-exchange membrane as electrolyte, powered the Gemini spacecraft, and two versions of the intermediate-temperature modified Bacon cell powered the Apollo vehicle and

TABLE 7 Battery-operated Motors

Hitachi standard model	Power supply	Voltage Operating range	Voltage Nominal	No load Speed, rpm	No load Current, amp	At maximum efficiency Speed, rpm	At maximum efficiency Current, amp	At maximum efficiency Torque oz-in.	At maximum efficiency Torque g-cm	At maximum efficiency Output, w	At maximum efficiency Efficiency, %	At maximum output power Speed, rpm	At maximum output power Current, amp	At maximum output power Torque oz-in.	At maximum output power Torque g-cm	At maximum output power Output, w	At maximum output power Efficiency, %	Starting torque oz-in.	Starting torque g-cm
RE-14-2280	Dry cell	1.5-3.0	1.5	7,500	0.220	5,050	0.630	0.092	6.6	0.330	39.0	3,800	0.860	0.141	10.1	0.380	35.0	0.284	20.4
RE-26-2680	Dry cell	1.5-3.0	1.5	5,500	0.160	3,950	0.600	0.139	10.0	0.400	48.0	2,750	0.960	0.243	17.5	0.480	40.0	0.487	35.0
RE-26L-22110	Dry cell	1.5-3.0	3.0	6,800	0.120	5,400	0.450	0.167	12.0	0.650	50.0	3,500	0.900	0.382	27.5	0.950	40.0	0.765	55.0
RE-36-24110	Dry cell	1.5-4.5	3.0	5,700	0.130	4,400	0.500	0.250	18.0	0.840	57.5	2,800	1.000	0.584	42.0	1.090	44.5	1.168	84.0
RE-56-30110	Dry cell	3.0-6.0	4.5	7,300	0.150	5,800	0.800	0.417	30.0	1.750	54.5	3,750	1.700	1.000	72.0	2.700	44.0	2.029	146.0
RE-56L-30110	Dry cell	3.0-6.0	4.5	5,400	0.100	4,400	0.580	0.445	32.0	1.430	58.0	2,700	1.500	1.307	94.0	2.500	45.0	2.585	186.0
FM-13-16130	Dry cell	1.5-3.0	1.5	4,500	0.130	2,900	0.300	0.053	3.8	0.110	25.5	2,300	0.370	0.072	5.2	0.120	25.5	0.146	10.5
FM-25-18150	Dry cell	1.5-3.0	3.0	8,100	0.100	6,000	0.380	0.104	7.5	0.450	41.0	4,100	0.640	0.201	14.5	0.590	34.0	0.403	29.0
RM-36-3260	Dry cell	1.5	1.5	6,300	0.195	4,850	0.700	0.139	10.0	0.490	53.5	3,100	1.300	0.306	22.0	0.680	44.5	0.612	44.0
RM-36-24110	Dry cell	1.5-3.0	3.0	7,200	0.100	5,800	0.440	0.153	11.0	0.630	51.0	3,600	1.000	0.396	28.5	1.030	40.5	0.792	57.0
FT-13FO-14170	Power pack	6.0-12.0	12.0	19,200	0.200	14,200	0.560	0.166	12.0	1.700	27.5	10,000	0.900	0.320	23.0	2.300	23.0	0.640	46.0
FT-13UO-16130	Power pack	6.0-12.0	12.0	32,000	0.470	22,400	1.000	0.208	15.0	3.400	32.0	16,000	1.380	0.347	25.0	4.000	28.0	0.690	50.0
FT-16-16130	Power pack	6.0-12.0	12.0	27,400	0.320	19,800	0.880	0.278	20.0	4.000	42.0	13,800	1.300	0.493	35.5	4.900	36.0	1.000	72.0
FT-16D-18110	Power pack	6.0-12.0	12.0	37,000	0.470	25,000	1.400	0.347	25.0	6.250	45.0	18,600	1.940	0.541	39.0	7.250	41.0	1.083	78.0
FT-36-22120	Power pack	6.0-12.0	12.0	20,000	0.400	14,000	1.500	0.639	46.0	6.400	44.0	10,000	2.200	1.042	75.0	7.500	39.5	2.083	150.0
FT-36D-24110	Power pack	6.0-12.0	12.0	23,600	0.450	16,200	1.700	0.667	48.0	7.800	48.0	12,000	2.500	1.069	77.0	9.200	44.0	2.139	154.0
RM-33-18200	Dry cell	1.5-3.0	1.5	2,850	0.050	2,100	0.150	0.056	4.0	0.084	39.0	1,400	0.230	0.106	7.6	0.106	32.0	0.211	15.2
FM-36-18200	Dry cell	1.5-3.0	1.5	3,200	0.030	2,500	0.125	0.044	3.2	0.080	43.5	1,600	0.260	0.107	7.7	0.124	32.5	0.214	15.4
FM-36K-08700	Lead battery	6.0-12.0	12.0	4,250	0.015	3,200	0.050	0.125	9.0	0.288	47.5	2,100	0.095	0.278	20.0	0.420	38.0	0.556	40.0
FM-65-18350	Dry cell	1.5-3.0	1.5	1,730	0.025	1,250	0.090	0.069	5.0	0.062	46.0	850	0.150	0.128	9.2	0.080	39.0	0.257	18.5
RM-75K-22250	Dry cell	1.5-4.5	3.0	3,100	0.050	2,550	0.200	0.181	13.0	0.331	55.0	1,500	0.490	0.521	37.5	0.560	41.5	1.000	72.0
RS-34S-3270	Dry cell	1.5-3.0	1.50	5,600	0.300	3,900	1.000	0.174	12.5	0.480	39.0	2,800	1.400	0.278	20.0	0.560	35.0	0.569	41.0
RS-34-4535	Nickel cadmium	1.25-2.50	1.25	6,400	0.830	4,900	2.500	0.306	22.0	1.060	34.0	3,300	4.150	0.611	44.0	1.460	28.5	1.264	91.0
RS-34-4535	Nickel cadmium	1.25-2.50	2.50	11,000	0.850	8,900	3.500	0.486	35.0	3.500	36.5	5,600	7.800	1.278	92.0	5.200	26.0	2.569	185.0
RS-54K-5550	Nickel cadmium	3.0-6.0	3.60	6,600	0.700	5,600	3.000	1.250	90.0	5.100	46.5	3,300	9.000	4.305	310.0	10.100	32.0	8.680	625.0
RS-54K-38100	Lead battery	6.0-12.0	12.00	11,000	0.450	9,300	2.700	2.361	170.0	15.500	48.0	5,500	7.800	7.638	550.0	31.000	32.5	15.277	1,100.0
RS-85-42120	Lead battery	6.0-24.0	12.00	7,200	0.400	5,850	2.550	4.166	300.0	17.500	56.0	3,600	6.200	11.110	800.0	28.800	39.0	22.221	1,600.0

its Lunar Excursion Module. These developments represented the first functional uses of fuel cells.

In fiscal year 1964, OART spent about $1.8 million on fuel cell projects ranging from basic research to prototype development.

Research on high-performance thin electrodes that promise drastic cuts in fuel-cell weight and volume should benefit both earth and space applications. Progress has reduced the ratio from about 150 lb/kw to about 70 lb/kw, exclusive of fuel and fuel tankage; 30 to 40 lb/kw for fuel cell plus auxiliaries now appears to be a reasonable expectation.

Primary fuel cells are those through which reactants are passed only once. They are useful in space for only limited periods because the product of power and duration (= energy) determines the amount of fuel and oxidant that must be carried aloft. For extended missions, therefore, other primary sources of energy must be used. In connection with solar and nuclear energy sources and conversion devices, fuel cells may be used for energy storage, as secondary power sources during darkness (solar primary power), during emergencies, and during periods of peak power demands. Among the methods of regenerating reactants from products, only electrolysis and thermal treatment have shown promise. Even so, it is not yet clear whether regenerative fuel cells will be competitive with secondary batteries or other secondary conversion devices.

What do we expect from space-type fuel cells? Immediate, prime considerations are high power density and reliability. The Gemini and Apollo fuel cells, for example, were one-sixth to one-tenth the weight of the best available primary batteries that are capable of delivering the same total amount of energy. Furthermore, the water product, an additional bonus not available from batteries, was used by the astronauts.

Most of the discussion in this volume will concern the civilian "workhorses," the zinc-carbon dry cell and the lead–sulfuric acid automobile battery. Supplementing these are the zinc–alkaline–manganese dioxide units, the mercury cells, and the alkaline-nickel-iron, as well as the nickel-cadmium secondary cells. There will be discussions of the "air cell," the fuel cells, the silver units, and the water-activatable units, which have some civilian applications but are primarily used by the military, and then of some exotic units entirely of a military, research, and noneconomic nature.

Small motors are made in quantities of millions per day and these are battery powered, as shown in Table 7.

REFERENCE

1. Ernst M. Cohn, NASA's Fuel Cell Program, in Robert E. Gould (ed.), "Fuel Cell Systems," Advances in Chemistry Series, no. 47, pp. 1–8, ACS, Washington, D.C., 1965.

chapter 4

The Workhorse—Dry Cells—the Zinc Ammonium Chloride | Manganese Dioxide | Carbon System

The dry cell is the outgrowth of the Leclanché cell. It was desired to make its electrolyte unspillable. Absorbents and fillers, including sand, sawdust, cellulose, asbestos fiber, plaster of paris, and spun glass, were tried. In 1888 a successful dry cell was produced by Gassner. It consisted of a zinc can serving as anode and as the cell container, a carbon rod surrounded by the depolarizing mixture which was wrapped in cloth, and the electrolyte in the form of a jelly. Its voltage on open circuit was about 1.3 v, and its short-circuit current about 6 amp.

The reaction of the Lechlanché and the dry cells may be given as

$$Zn + 2NH_4Cl + 2MnO_2 \rightleftharpoons Zn(NH_3)_2Cl_2 + H_2O + Mn_2O_3$$

Manufacture

Zinc of high purity serves as a container for the cell. The electrolyte is an aqueous solution of NH_4Cl and $ZnCl_2$. It is held partly in an absorbent material and partly in the mixture of ground carbon and MnO_2. The lining separating the zinc and the depolarizing mixture must allow electrolytic but not metallic conduction, in that the latter would cause an internal short-circuit. The depolarizing mixture is bulky and occupies most of the interior of the cell. The electrolyte is sometimes made into a jelly with colloidal materials such as gum tragacanth, agar, gelatin, flour, or starch. The electrolyte, therefore, will not spill, whether it be completely sealed over at the top, the method common in American practice, or provided with a gas vent.

When the cell is new, the surface of the composite carbon-MnO_2 electrode may be taken as the outside surface of this mixture next to the zinc. As the cell is discharged, the MnO_2 is reduced, and the effective surface of the electrode moves toward the carbon rod, which is in the center axially with the cell. This carbon rod serves only to conduct the current out of the mixture to the terminal.

Large sizes are for ignition, signaling, and miscellaneous intermittent use, and smaller sizes for flashlights, radio batteries, and similar purposes. In the larger size, sheet zinc is the container, the bottom being soldered with lap seams, while zinc stampings are employed for the smaller cells.

The conductivity of the MnO_2 depolarizer is low so that granulated carbon, more or less completely graphitized, or acetylene black is added to lower resistance. The NH_4Cl electrolyte must be free of metals such as Cu, Pb, Fe, As, Ni, Co, and Sb, which may be plated out by the zinc, as well as free from negative radicals, such as sulfates, which form compounds less soluble than the chlorides. Insulation and sealing compounds are usually resin, with waxes or bituminous pitches with fillers such as ground silica, fibrous talc, and coloring matter added. The cell is insulated by a strawboard container or jacket surrounding it.

The MnO_2 pyrolusite is a refined ore, or may be produced by anodic oxidation. Its efficiency depends upon the percentage of MnO_2 and its state of hydration, and its crystalline state as indicated by x-ray examination. The ore must be free from Cu, Ni, or Co. Iron to the extent of 1% is not unusual.

An increase in grain size reduces the internal resistance of the carbon-MnO_2 mixture, while a decrease increases the depolarizing power per unit weight of MnO_2. A high degree of porosity is desirable inasmuch as the depolarizing power depends upon the surface area.

The zinc container is lined with a sulfite and ground wood-pulp board at the sides and bottom, into which the MnO_2-NH_4Cl-$ZnCl_2$ mixture is tamped around the central carbon. The pulpboard lining is then folded down over the top of the mixture and the cell is sealed with a sealing compound. The construction is shown in Fig. 10. In a modified form the depolarizer is extruded to shape.

A second method is the bag or sack type, in which the carbon rod with its mixture of electrolyte and depolarizer is wrapped in muslin, forming a unit which is placed in the zinc can, leaving space for the electrolyte as shown in Fig. 11. The solution of sal ammoniac and ZnCl₂ is thickened with gelatinous materials. Spacers to separate the bag from the zinc can are desirable. This form is for small flashlight batteries, increasing

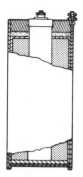

Fig. 10 Section of paper-lined cell.

Fig. 11 Section of bag-type cell.

the life of small cells, which may be shorter than that of the larger sizes even when standing on open circuit.

In an electrolyte containing more than 8 to 10% NH_4Cl the diamine may be formed, whereas for lower concentrations an oxychloride may be the principal reaction product:

$$4Zn + 8MnO_2 + 8H_2O + ZnCl_2 \rightleftharpoons 4Mn_2O_3 \cdot H_2O + ZnCl_2 \cdot 4Zn(OH)_2$$

The choice of reaction is dictated by factors such as diffusion effects, severity of service, degree of exhaustion of the cell system, and the like, and the transition from the first to the second can probably occur without inflection in operating voltage.[1]

Copeland and Griffith[2] established the formation of $ZnO \cdot Mn_2O_3$ (hetaerolite) as a possible cathodic product. The overall cell reaction may be written as

$$Zn + 2MnO_2 \rightleftharpoons ZnO \cdot Mn_2O_3$$

in which no ammonium chloride is removed from the cell system.

The anode reaction is simple. Even during continuous discharge, polarization or changes in concentration and pH of the anolyte have slight effect on zinc potential,[3] perhaps 0.2 v at the current drain of flashlight service.

Mercury is generally introduced as $HgCl_2$ dissolved in the electrolyte, seldom exceeding 0.25% because of the embrittlement and loss of strength of the amalgamated metal. From the electrolyte it is deposited on the zinc by replacement.

The wasteful corrosion of zinc in cells on long-time intermittent service may be as much as 50% of the weight of metal generating current.

Aluminum, with its high electrochemical equivalent (2.98 ah/g, 2.2 for magnesium and 0.82 for zinc), would be attractive were it not for its high corrosion rate. It shows no voltage advantage over zinc. Its stability depends on an oxide film and its true potential is not realized.

The energy of an 85-g D cell is equal to almost 13,000 joules and will deliver about 2.73 ah before its voltage falls to 1.13 v on a discharge through 83.3 ohms for 4 hr/day. The average voltage is 1.3 v.

Table 8 gives a range of recommended sizes of zinc-carbon cylindrical cells and Table 9 of flat cells, with the corresponding designations of International Electrotechnical Commission, American National Standards Institute, and Bureau of Standards committees. The No. 6 general-purpose cell shows an initial short-circuit amperage of 43 amp, a deterioration of 7% in 6 months, a life of 77 hr in the heavy intermittent-service test and 250 days in the light intermittent test. The small CD hearing-aid batteries have a 50-hr life in the heavy earphone-service test.

Ramsey[4] described a 1,000-v battery for Geiger counters for unmanned balloons which measure radiation in the stratosphere. The potential could not fall more than 0.1% per week.

Wilke[5] incorporated lithium chloride in the liquid electrolyte to operate at $-40°C$ ($-40°F$). Cells have been developed and produced which at

TABLE 8 Sizes of Zinc-Carbon Cylindrical Cells

Cell designations	IEC designations	Nominal dimensions				Approximate volume		Approximate weight	
		Diameter		Can height					
		in.	mm	in.	mm	cu in.	cu cm	lb	g
0	....	0.45	11.4	0.13	3.3	0.02	0.3	0.002	0.9
N	R-1	0.44	11.2	1.06	26.9	0.16	2.6	0.012	5.4
AAA	R-03	0.39	9.9	1.69	42.9	0.20	3.3	0.018	8.2
R	R-4	0.53	13.5	1.31	33.3	0.29	4.8	0.023	10.4
AA	R-6	0.53	13.5	1.88	47.8	0.41	6.7	0.033	15
A	R-8	0.63	16.0	1.88	47.8	0.58	9.5	0.046	21
B	R-12	0.75	19.1	2.13	54.1	0.94	15.4	0.077	35
C	R-14	0.94	23.9	1.81	46.0	1.26	20.6	0.10	45
D	R-20	1.25	31.8	2.25	57.2	2.76	45.2	0.22	100
E	R-22	1.25	31.8	2.88	73.2	3.53	57.8	0.29	132
F	R-25	1.25	31.8	3.44	87.4	4.22	69.2	0.35	159
G	R-26	1.25	31.8	4.00	101.6	4.91	80.5	0.40	181
J	R-27	1.25	31.8	5.88	149.4	7.2	118	0.6	272
#6	R-40	2.50	63.5	6.00	152.4	29.5	483	2.2	998

TABLE 9   Sizes of Zinc-Carbon Flat Cells

Cell designations	IEC designations	Nominal dimensions						Approximate cross-sectional area	
		Length		Width		Thickness			
		in.	mm	in.	mm	in.	mm	sq in.	sq cm
F12	.....	.....	.....	0.63	16.0	0.28	7.1	0.31	2.0
F15	F15	0.56	14.2	0.56	14.2	0.12	3.0	0.31	2.0
F17	.....	.....	.....	0.72	18.3	0.29	7.4	0.41	2.6
F20	F20	0.94	23.9	0.53	13.5	0.11	2.8	0.50	3.2
F22	F22	0.94	23.9	0.53	13.5	0.28	7.1	0.50	3.2
F24	.....	.....	.....	0.91	23.1	0.26	6.6	0.65	4.2
F25	F25	0.89	22.6	0.89	22.6	0.23	5.8	0.79	5.1
F30	F30	1.25	31.8	0.84	21.3	0.13	3.3	1.05	6.8
F40	F40	1.25	31.8	0.84	21.3	0.21	5.3	1.05	6.8
F45	.....	1.50	38.1	0.84	21.3	0.11	2.8	1.26	8.1
F50	F50	1.25	31.8	1.25	31.8	0.14	3.6	1.56	10.1
F60	F60	1.25	31.8	1.25	31.8	0.15	3.8	1.56	10.1
F70	F70	1.70	43.2	1.70	43.2	0.22	5.6	2.89	18.6
F80	F80	1.69	42.9	1.69	42.9	0.25	6.4	2.86	18.5
F90	F90	1.69	42.9	1.69	42.9	0.31	7.9	2.86	18.5
F96	.....	2.13	54.1	1.75	44.5	0.21	5.3	3.73	24.1
F100	F100	2.38	60.5	1.78	45.2	0.41	10.4	4.24	27.4

−40°C give capacities of 10 to 20% of those obtained at 21°C (70°F), depending on the severity of the drain.

When a cell is gradually cooled to a low temperature, its internal resistance gradually rises for a time and at a critical temperature rises abruptly, corresponding to the electrolyte solidification. Rise of internal resistance prevents operation of the unit.

The current is carried both by the electrolyte electronically and by the conductive depolarizing mix and is little affected by temperature. The conduction through the electrolyte is ionic and temperature dependent, affecting viscosity of the electrolyte and ionic mobility.

The electrolyte can be maintained in a liquid state by antifreeze agents such as alcohol or ethylene glycol, but the conductivity of such solutions at −40°C is low. Salts such as $ZnCl_2$ and $CaCl_2$ may be added to reduce solidification or congealing of the electrolyte at −40°C, but the viscosity and resistance remain high. Solutions containing lithium chloride are fluid, and the addition of the LiCl to the electrolyte improves its conductivity at −40°C. These solutions show no detrimental effect on starch and do not affect shelf life adversely.

Zinc-carbon batteries have been developed in several hundred styles, in various voltages, shapes, and sizes, with numerous terminal arrangements. Types are available in voltages from 1.5 to 510 v connected in series for higher voltages, in parallel for greater service capacity, or in series-parallel to give higher voltage and service capacity.

The zinc-carbon type is the most widely used system. Three categories

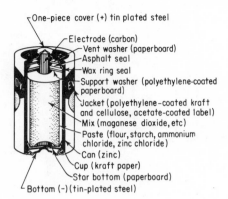

Fig. 12 Cross section of standard round cell. (*Eveready, Union Carbide Corporation.*)

exist: (1) round—as unit cells or in assembled batteries, (2) flat, as shown in Table 9, and (3) wafer or multicell batteries. The difference is physical; the flat cell arranges these materials in a laminated structure while in the round cell they are concentric (see Figs. 12, 13, and 14).

Wafer-type cells are rectangular in shape with the corners slightly rounded. They consist of a sandwich of artificial manganese dioxide mix

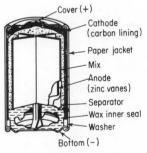

Fig. 13 Cross section of external cathode or round cell. (*Eveready, Union Carbide Corporation.*)

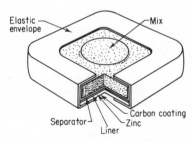

Fig. 14 Cross section of flat cell. (*Eveready, Union Carbide Corporation.*)

between disks of flat zinc and carbon electrodes. The sandwich is wrapped in pliofilm envelopes and sealed. At this point the individual cells may be tested for quality control before they are assembled in series into units.

A spot of silver wax on the positive and negative sides of the cells provides electrical contact between cells as they are stacked together, thus eliminating open-circuit hazards of pin-wire connectors or unreliable pressure contacts. The cell stack is then wrapped in Mylar film and is ready for packaging as a finished battery.

Battery Service Life

The ampere-hour capacity of a dry battery is not a fixed value. It varies with current drain, operating schedule, cutoff voltage, temperature, and storage of the battery prior to use. Service capacity for dry cells is given in Table 10 for three different current drains. The values are for fresh batteries at 70°F; the operating schedule is 2 hr/day. The cutoff voltage is 1.0 v per 1.5-v cell for the first 10 cells listed and 0.8 v per 1.5-v cell for all of the remaining cells. The data are based on starting drains and constant resistance tests. Service capacity is shown for single 1.5-v cells.

An external cathode construction is shown in Fig. 13. An impervious, inert carbon wall is the container and current collector. Zinc vanes are inside. In flat cells, carbon is coated on a zinc plate to form a duplex electrode—a combination of the zinc of one cell and the carbon of the adjacent one. The flat cell (Fig. 14) contains no expansion chambers or carbon rod. This increases the amount of depolarizing mix available per unit cell volume and therefore the energy content. The energy-to-volume ratio of a battery utilizing round cells is inherently poor because of the voids occurring between cells. These factors account for an energy-to-volume improvement of nearly 100% for flat cells compared to round cell assemblies. Resistance to leakage is better in flat battery construction.

Quality Control and Testing

The closed-circuit voltage or working voltage of a zinc-carbon cell falls gradually as it is discharged (see Fig. 15). The service hours delivered are greater as the cutoff or end-point voltage is lower. Cutoffs range

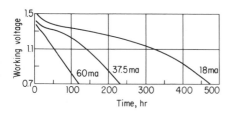

Fig. 15 Voltage discharge characteristics of carbon-zinc D-size battery discharged 4 hr/day. (*Eveready*, *Union Carbide Corporation*.)

TABLE 10 Service Capacities

Cell	Starting drain, ma	Service capacity, hr	Cell	Starting drain, ma	Service capacity, hr	Cell	Starting drain, ma	Service capacity, hr
N	1.5	275	105	0.4	210	143	2	190
	7.5	52		2	30		10	40
	15	24		4	8		20	15
AAA	2	290	108	0.5	435	145	2	510
	10	45		2.5	103		10	105
	20	17		5	51		20	43
AA	3	350	109	0.6	710	146	2	560
	15	40		3	155		10	115
	30	15		6	75		20	50
B	5	420	112	0.7	210	148	2	610
	25	65		3.5	35		10	150
	50	25		7	12		20	60
C	5	430	114	0.7	300	162	3	550
	25	100		3.5	57		15	150
	50	40		7	25		30	65
D	10	500	117	0.8	475	163	3	600
	50	105		4	98		15	165
	100	45		8	49		30	72
E	15	400	125	1	500	165	3	770
	75	70		5	105		15	200
	150	30		10	45		30	90
F	15	520	127	1	475	172	5	780
	75	105		5	150		25	200
	150	45		10	72		50	90
G	15	820	132	1.3	275	175	5	1,000
	75	150		6.5	40		25	260
	150	65		13	16		50	110
6	50	700	133	1.3	450	176	10	910
	250	150		6.5	80		50	165
	500	70		13	35		100	63
104	0.4	200	135	1.3	450			
	2	28		6.5	108			
	4	7		13	52			

SOURCE: Eveready, Union Carbide Corporation.

TABLE 11 Performance on Standard Tests

Test name	Daily discharge schedule	Resistance per 1½-v unit, ohms	End point per 1½-v unit, v	Cell size	American National Standards Institute (ANSI) initial requirement
General-purpose flashlight to represent 0.5-amp lamp	One 5-min period	2.25	0.65	D (General purpose)	400 min
General-purpose flashlight to represent 0.3-amp lamp	One 5-min period	4	0.75	C AA	325 min 80 min
General-purpose flashlight to represent 0.22-amp lamp	One 5-min period	5	0.75	AAA	50 min
Heavy-industrial flashlight to represent 0.3-amp lamp	32 4-min periods	4	0.9	D (Industrial)	800 min
Light-industrial flashlight to represent 0.3-amp lamp	Eight 4-min periods	4	0.9	D (General purpose) D (Industrial)	600 min 950 min
Railroad lantern to represent 0.15-amp lamp	Eight ½-hr periods	8	0.9	F	45 hr
Photoflash test	60 1-sec periods (1 discharge per min for 1 hr/day)	0.15	0.5 D size 0.25 C and AA sizes	D (Photoflash) C (Photoflash) AA (Photoflash)	800 sec 700 sec 150 sec
Radio A	One 4-hr period	25	1.0	F G	140 hr 170 hr
Radio B	One 4-hr period	166.67	1.0	F40 A F90	30 hr 130 hr 225 hr
Heavy, intermittent	Two 1-hr periods	2.67	0.85	F (4 par) #6 (General purpose) #6 (Industrial)	70 hr 70 hr 100 hr

from 0.8 to 1.1 v per cell, depending upon the application. The cutoff should be made as low as possible so as to use the available energy in the battery.

Performance on standard tests is shown in Table 11.

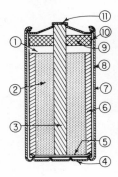

Fig. 16 Cross section of a typical cylindrical carbon-zinc dry cell. (1) Air space. (2) Manganese dioxide–carbon–electrolyte cathode mix. (3) Carbon electrode. (4) Steel bottom. (5) Paper bottom washer. (6) Starch-flour-electrolyte gel separator. (7) Steel jacket. (8) Laminated paper tube. (9) Pitch seal. (10) Zinc anode can. (11) Steel top. (*Radio Corporation of America.*)

Service capacity ranges from several hundred milliampere-hours (mah) to 30 ah. When assembled into batteries, the capacities range to over 100 ah.

TABLE 12 Typical Cathode Mix and Electrolyte Composition of a Carbon-Zinc Dry Cell

Material	Composition, %
Typical black mix	
Manganese dioxide	62
Acetylene black*	8
Zinc chloride	14
Sal ammoniac	1
Water	15
	100
Typical electrolyte	
Ammonium chloride	9
Zinc chloride	26
Water	65
	100

* Some mixes use graphite instead of acetylene black.

The chemical efficiency of a zinc-carbon battery improves as current density decreases. Over a range of current density, service life may be tripled by halving the current drain.

In the RCA version of the dry cell, a typical cathode mix is given in Table 12, and the construction with an outer steel can is shown in Fig. 16.

Sizes and Application

Four different classes of zinc-carbon cells are generally available from most battery manufacturers. The four general areas of application for these different classes are: radio, general-purpose, flashlight and photoflash, and heavy-duty industrial. The dry cells vary in their chemical composition depending on the application for which they are intended. Thus, a dry cell or battery intended for radio applications contains a higher percentage of active electrochemical materials than a dry cell or battery intended for photoflash applications. The higher percentage of electrochemical materials increases the overall capacity of the cell, enabling it to remain in service longer than a similar-size cell which is designed for photoflash applications. Conversely, a dry cell or battery intended for photoflash applications contains a higher percentage of carbon than the cell intended for radio applications. The higher carbon content enables the cell to deliver the high current of short duration necessary to fire the flash bulb. Although either of these classes of dry cell or battery will operate in either application, the most satisfactory results are obtained when each class of dry cell or battery is used in the application for which it is specifically designed.

The service capacity depends on the relative time of discharge and recuperation periods. The performance is normally better when the service is intermittent. Continuous use is not necessarily inefficient if the current drain be light. Figure 17 illustrates the service advantage to be obtained by proper selection of a battery for an application.

Zinc-carbon batteries are designed to operate at 70°F. The higher

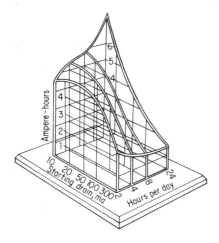

Fig. 17 Battery service life as a function of initial current drain and duty cycle (D-size zinc-carbon cell). (*Eveready, Union Carbide Corporation.*)

the battery temperature during discharge, the greater the energy output. High temperature reduces shelf life. Prolonged exposure to temperatures above 125°F causes rapid disintegration.

Shelf life is the time at 70°F after which a given battery retains a specified percentage (usually 90%) of its original energy content. Shelf life is reduced by high temperatures because of zinc corrosion, side chemical reactions within the cells, and moisture loss through evaporation. The shelf life of a battery stored at 90°F is about one-third that of one stored at 70°F.

Service life at low temperatures is reduced because of decreased chemical activity in the cell. The effects are more pronounced for heavy than for light drains. When a zinc-carbon battery has reached a temperature of 0°F, there is little service except at light drains. Since a battery does not reach the temperature of its surroundings immediately, insulation is helpful.

To appreciate the position batteries occupy in our civilization, a tabulation has been made of the important commercial forms employed by their designer, the manufacturer of electronic and electrical components, by tabulating the batteries commonly warehoused, distributed, and sold by the National Electronic Dealers Association (NEDA). This distribution might be considered a measure of possible popularity of acceptance by consumers and might allow an evaluation of battery sizes, performance, and application. NEDA in its merchandising lists only comparable number designations of competitive battery makers. The tables have been developed to be a source of battery data, as well as giving an insight of the competition between systems such as zinc-carbon, alkaline, mercury, and silver as represented by the demands on the distributors and warehousers. Table 13 gives details on A batteries. Note the 24A is a competitive alkaline unit. Table 14 deals with commercial B batteries called for by industry. Table 15 represents popular A-B batteries with ranges of voltages, all high by the ordinary standards. All the items are assemblages of zinc-carbon cells in a wide variety of electrical connections, terminals, etc. Table 16 gives units for flashlights, test equipment, portable items, and heavy-duty B applications. The figures in Table 17 are for ignition, flashers, lighting, lanterns; there is also an alkaline unit among them. Table 18 covers radios, transistors, instruments, portable units, transceivers; in this group the demand is heavy for the zinc-carbon type—the alkaline, mercury, and silver units find application in a competitive manner. Table 19 is an extension of Table 18 for similar purposes and further shows the competition.

However, from Tables 13 to 19 the major demands by industry, as represented by the type of stock carried by the warehousers and distributors, rely on the zinc-carbon system.

TABLE 13 Characteristics of A Batteries

NEDA	Designation	ANSI desig-nation	Volt-age	Cur-rent, ma	Diameter in.	Diameter mm	Number of cells and sizes	Length in.	Length mm	Width in.	Width mm	Overall height in.	Overall height mm	Weight oz	Volume cu in.	Volume cu cm	Other designations
1	Low voltage	.....	6	500	.....	.....	8/F	3.906	99.21	2.750	69.85	5.437	138.09	36	58.5	913.77	BA2031J
2	Low voltage	.....	6	25	.....	.....	4/AA	1.219	30.96	1.219	30.96	2.343	59.51	2.5	3.4	53.10	
3	Low voltage	3F	4.5	250	.....	.....	3/F	4.000	101.60	1.437	36.50	4.125	104.78	.....	20.7	323.33	7D7003, Z736
4	Low voltage	F4D	1.5	1,000	.....	.....	4/F	2.625	66.68	2.625	66.68	4.125	104.78	22	26.5	413.93	BA65, Z94
5	Low voltage	F6D	1.5	1,500	.....	.....	6/F	3.937	100.00	2.750	69.85	4.125	104.78	31	40.5	632.61	Z96
6	Low voltage	4FD	6.0	250	.....	.....	4/F	2.687	68.25	2.687	68.25	4.062	103.17	22	26.5	413.93	
7	Low voltage	3G	4.5	300	.....	.....	3/G	4.000	101.60	1.437	36.50	4.750	120.65	19	24	374.88	BA2261J, 17-266, Z83A
8	Low voltage	.....	7.5	50	.....	.....	5/B	3.906	99.21	0.828	21.03	2.843	72.21	9	9.21	143.86	
9	Low voltage	.....	7.5	70	.....	.....	5/172	2.156	54.76	1.937	49.19	3.031	76.98	8	13	203.06	Z750, 17-275, 7D7009
11	Low voltage	F2	1.5	500	.....	.....	2/F	2.625	66.68	1.375	34.93	4.250	107.95	12	14.8	231.17	
13	Transistor radio	.....	1.5	150	1.343	34.11	1/D	.....	.....	.....	.....	2.406	61.11	3	3.19	49.82	13267, 95MW
14	Transistor radio	.....	1.5	80	1.031	26.18	1/C	.....	.....	.....	.....	1.937	49.20	1.4	1.25	19.52	Z7, 6446, 17-222 EX5750-3
15	Transistor pocket radio	.....	1.5	25	0.563	14.30	1/AA	.....	.....	.....	.....	1.968	49.98	0.6	0.48	7.49	Z8, 6447, EX5700-4 17-321
16	Low voltage	4F25	6	.....	.....	.....	.....	3.937	100.00	1.437	36.50	10.875	276.23	42	54.8	855.97	
17	Low voltage	F8d	1.5	2,000	.....	.....	8/F	3.937	100.00	2.750	69.85	5.500	139.79	6.5	10.8	168.70	
18	Low voltage	.....	1.5	300	.....	.....	2/D	2.781	70.63	1.406	35.71	3.031	76.98	10.5	15.2	237.42	
19	Low voltage	3D	4.5	150	.....	.....	3/D	4.000	101.60	1.375	34.93	3.000	76.20	5.9	4.92	62.48	Z450
20	Low voltage	.....	1.5	300	1.343	34.11	1/G	.....	.....	.....	.....	4.156	105.56	5	5.5	85.91	Z5, 7D8020 17-272
23	Low voltage	.....	1.5	250	1.343	34.11	1/F	.....	.....	.....	.....	4.062	103.17	0.3	0.2	3.12	BA 231/U, Z1
24P	Photoflash	.....	1.5	20	0.406	10.31	1/AAA	.....	.....	.....	.....	1.75	44.45	0.4	1.6	24.99	
24A	Alkaline	L30	1.5	20	0.400	10.16	1/AAA	.....	.....	.....	.....	1.673	42.49				
26	Low voltage	.....	7.5	30	.....	.....	5/175	2.562	65.07	2.031	51.58	2.812	71.42	12	14.1	220.24	Z707

TABLE 14 Characteristics of B Batteries

NEDA	Designation	ANSI designation	Voltage	Current, ma	Number of cells and sizes	Length in.	Length mm	Width in.	Width mm	Overall height in.	Overall height mm	Weight, oz	Volume cu in.	Volume cu cm	Other designations
200	Portable	45N, 45F40	67.5	10	45/135	2.812	71.42	1.375	34.93	3.812	96.82	12	14.2	221.80	17-304, TD-7200, Z67
200C	Communications		67.5	15	45/135	2.812	71.42	1.375	34.93	3.812	96.82	12	14.2	221.80	
201	Portable	30F40	45	10	30/135	2.656	67.46	1.000	25.40	3.750	95.2	7.8	9.6	149.95	BA56, Z455
202	High voltage	30F90	45	40	30/165	3.625	92.08	1.844	46.84	5.625	142.88	30.0	36.4	568.56	BA59, Z783
203	High voltage	45F30	67.5	6	45/135	2.812	71.48	1.375	34.93	2.500	63.50	7.4	9.46	147.76	Z457
204	High voltage	60F40	90	10	60/135	3.719	94.46	1.375	34.93	3.750	96.25	15	18.8	293.65	Z490
205	High voltage	30F70	45	25	30/AA	3.125	79.38	2.375	60.33	4.187	106.35	19.0	28.6	446.73	BA63
206	High voltage	30F80	45	40	30/163	3.562	90.47	2.312	58.72	4.625	117.68	26.8	36.0	562.32	BA2231J, Z530
207	High voltage	30F96	45	70	30/172	4.375	111.13	2.688	68.28	5.500	139.70	50.0	53.4	834.10	
208	High voltage		15	2.5	10/112	1.031	26.18	0.625	15.87	1.453	36.90	0.95	0.95	14.83	Z11M
210	Radio paging	20F20	30	2.5	20/112	1.062	26.97	0.625	15.88	2.562	65.07	1.6	1.59	248.35	Z13M
211	Radio paging	45F25	67.5	8	44/125	1.922	48.82	1.016	25.81	5.562	141.27	8.6	9.96	155.57	6482 Z477
212	Radio paging		75	10	52/136	1.937	49.19	1.468	37.28	6.48	164.59	14.8	8.2	128.08	Z437
213	Pocket receiver		45	4	30/112	1.093	27.76	0.625	15.87	3.68	93.47	2.5	2.35	36.70	Z415
214		60F25	90	8	60/125	1.969	50.01	1.031	26.19	7.469	189.71	12.0	15.0	234.50	17-314, Z90
215	Pocket receiver	15F20	22.5	2.5	15/112	1.062	26.97	0.625	15.88	2.000	50.80	1.3	1.31	20.46	BA-2161J
216	High voltage		90	10	60/135	1.937	49.20	1.468	37.28	7.125	180.97	16	20.1	313.96	Z495
217	High voltage		67.5	3	46/114	1.328	33.73	0.984	24.99	3.50	88.90	4	4.45	69.50	17-305
220	High voltage	10F15	15	1.5	10/105	0.625	15.88	0.594	15.09	1.375	34.93	0.6	0.51	7.96	Z5M
221	High voltage	15F15	22.5	1.5	15/105	0.625	15.88	0.594	15.09	2.000	50.80	0.9	0.74	11.55	Z6M
222	High voltage	20F15S	30	1.5	20/105	0.625	15.88	0.594	15.09	2.609	66.27	0.9	0.971	15.15	
223	High voltage	20F15d	30	1.5	20/105	1.219	30.96	0.625	15.88	1.422	36.12	0.9	1.03	16.08	
224	High voltage		15	6	10/132	1.937	49.19	0.968	24.58	1.562	39.39	1.8	1.87	29.20	
225	High voltage	15F30	22.5	6	15/132	1.375	34.93	1.062	26.97	2.219	56.36	2.5	2.66	41.54	
226	High voltage	20F30	30	6	20/132	1.375	34.93	1.062	26.97	2.812	7.46	3.2	3.56	55.60	
227	High voltage		103.5	20	63/143	1.375	34.93	1.375	26.97	11.718	297.63	20	26.8	324.89	BA38

TABLE 15 Characteristics of A-B Batteries

NEDA	Designation	Voltage	Current, ma	Number of cells and size	Length		Width		Overall height		Weight, oz	Volume		Other designations
					in.	mm	in.	mm	in.	mm		cu in.	cu cm	
400	Portable	9/90	A50, B15	(6G)(60/165)	14.062	357.18	2.687	68.24	4.062	103.18	98	153	2,389.86	Z985
401	Portable	7.5/9/90	A50, B15	(6F)(80/162)	9.218	234.13	2.718	69.04	4.312	109.52	76	108	1,686.96	Z979, 6407
403	....	6/7.5/75	A50, B12	(5/176)(50/145)	8.562	217.47	3.750	95.25	2.437	61.89	57	78.5	1,226.17	
405	....	7.5/9/90	A50, B12	(6/176)(80/145)	8.875	225.42	2.125	56.26	3.781	96.03	46	71.1	1,110.58	
406	....	9/190	A50, B12	(6/F)(80/162)	9.218	234.13	2.718	69.00	4.327	109.90	76	106	1,655.72	Z909
410	....	1.5/90	A300, B14	(3/G)(60/146)	9.937	252.37	2.250	57.15	4.718	119.83	69	99	1,546.38	BA-48
413	Home receiver	1.5/90	A300, B12	(3/#6)(60/175)	15.697	398.45	2.218	56.337	7.812	198.42	260	452	7,060.24	Z802, ID7413, 17-325
415	Portable	7.5/9/90	A50, B12	(18F)(60/175)	7.844	199.24	4.125	104.77	9.875	250.82	236	320	4,998.40	
416	Portable	9/90	A50, B15	(6G)(60/165)	14.625	371.47	2.687	68.24	4.062	103.17	98	153	2,389.86	
425	Portable	1.5/90	A300, B14	(3/56)(60/135)	3.843	97.61	2.187	55.55	7.406	188.11	39	66.6	1,040.29	
428	....	9/90	A50, B12	(6/176)(60/145)	8.875	225.42	2.125	53.97	3.781	96.04	46	71.2	1,112.14	Z962
431	....	7.5/75	A50, B12	(5/176)(52/146)	8.375	212.72	3.625	92.07	2.25	57.15	44	68.4	1,068.40	Z775

TABLE 16 Characteristics of Flashlight and Heavy-duty B Batteries

NEDA	Designation	Type	ANSI designation	Voltage	Current, ma	Number of cells and sizes	Length in.	Length mm	Width in.	Width mm	Overall height in.	Overall height mm	Weight, oz	Volume cu in.	Volume cu cm	Other designations
700	Low voltage	Glo-plug ignition	....	1.5	500	2/F	2.625	66.67	1.937	49.20	4.031	102.38	12	12.4	193.6	BA-15A
701	Low voltage	Test equipment	....	3.0	250	2/AA	2.625	66.67	0.625	15.87	4.031	102.38	12	14.5	226.4	BA-205
702	High voltage	B and C	....	22.5	150	15/D	6.406	162.71	4.093	103.96	3.156	80.16	74	80.2	125.7	BA-8
703	Low voltage	Radio	....	3.0	500	4/F	2.625	66.67	2.625	66.67	2.656	67.46	23	28.1	438.9	BA-208
704	Low voltage	Flashlight	....	3.0	25	2/AA	1.218	30.93	0.625	15.87	2.656	67.46	2	1.62	25.3	
705	Low voltage	Flashlight	....	4.5	....		1.781	45.24	0.625	15.87	2.656	67.46	....	....	....	
706	Low voltage	Flashlight	....	4.5	50	3/B	2.437	61.90	0.843	21.41	3.062	77.77	....	....	....	BA-9
708	High voltage	Spec term	15B	22.5	50	15/B	4.250	107.95	2.625	66.68	3.312	84.12	24	291	4,545.4	BA-230
709	High voltage	Portable B—industrial	....	45	70	20/172	4.093	103.96	2.563	65.10	5.218	132.53	44	53	827.8	BA-36
710	High voltage	Screw B or C	15A-15F80	22.5	40	15/161	3.531	89.69	2.187	55.55	3.062	77.77	14	18.5	288.9	BA-2
711	High voltage	Hearing aid	....	45	20	30/146	3.062	77.77	1.875	47.62	4.968	126.18	14.5	25.8	402.9	BA-53
712	Low voltage	Portable C	....	4.5	150	3/D	4.062	103.17	1.406	35.71	3.031	76.98	14	18.1	282.7	BA-27
713	Low voltage	Flex wire-5 screw term	5B	7.5	50	5/B	4.250	107.95	0.937	23.80	3.312	84.12	9	8.24	128.7	
714	Low voltage	Screw term	3B	4.5	....		2.500	63.50	0.875	22.23	3.156	80.16	....	....	....	
715	High voltage	Heavy duty B	....	45	300	30/G	8.125	206.37	4.437	112.70	7.687	195.25	187	258.0	4,029.9	
716	High voltage	Heavy duty B	....	45	250	30/F	8.062	204.77	4.062	103.17	7.625	193.68	165	230.0	3,592.6	
717	High voltage	ASA VII term	30D	45	150	30/D	8.250	209.55	3.312	84.12	7.625	193.68	119.5	168.0	2,624.1	
718	High voltage	ASA XI term	3D	45	150	3/D	4.062	103.17	1.500	38.14	3.062	77.77	13	16.5	257.7	
719	Low voltage-C	Receiver A	....	1.5	4,500	18/F	4.593	116.66	3.875	98.42	7.750	196.85	....	....	....	
721	High voltage	ASA XII term	15B	22.5	....		4.250	107.95	2.625	66.68	3.062	77.77	....	....	....	BA-211/1J
722	High voltage Geiger	ASA XVIII term	200F20	300	2.5	200/112	2.687	68.25	2.219	56.36	3.906	99.21	14.5	23.4	365.5	BA-291/1J
723	High voltage	Industrial B	....	150	45	30/F	8.062	204.77	2.937	74.60	7.625	193.68	119.5	168.0	2,624.1	
726	High voltage	Industrial B	....	250	45	30/F	8.031	203.98	4.062	103.18	7.062	179.37	165	230.0	3,592.6	
727	High voltage	Industrial B	....	300	45	30/G	8.125	206.37	4.187	106.34	7.687	195.25	190	258.0	4,029.9	

TABLE 17 Characteristics of Ignition, Flasher, and Lantern Batteries

NEDA	Designation	Type	ANSI designation	Voltage	Current, ma	Diameter in.	Diameter mm	Number of cells and sizes	Length in.	Length mm	Width in.	Width mm	Overall height in.	Overall height mm	Weight, oz	Volume cu in.	Volume cu cm	Other designations
900	Low voltage	Instruments		1.5	1,000			4/F	2.625	66.67	2.625	66.67	4.312	109.52	23	26.5	413.9	BA-35, 2259
901	Low voltage	Telephone		3	1,000			8/F	3.781	96.04	2.687	68.24	5.812	147.62	44	54.6	852.8	BA-225 IJ
902	Low voltage	Emergency lighting		6				16/F	8.312	211.12	2.812	71.42	6.437	163.50	91	137.0	2,139.9	BA-222 IJ
903	Low voltage	Emergency lighting		7.5	1,000			20/F	7.25	184.15	4.06	103.12	6.437	163.50	122	173.0	2,702.3	
904	Low voltage	Emergency lighting		9	1,000			24/F	8.578	217.88	4.06	103.12	6.437	163.50	136	205	3,202.1	BA-207 IJ
905	Low voltage	Ignition, general		1.5	1,500	2.625	66.67	1×6					6.656	169.06	33	29.3	457.6	17-411, EX-5580-9, 13,250, 6MW, 4669S, 7D7905
905FC	Low voltage	Ignition, general		1.5	1,500	2.625	66.67	1×6					6.656	169.06	34	29.3	457.6	A-669C
906	Low voltage	Ignition, general		1.5	1,500	2.625	66.67	1×6					6.656	169.06	34	29.3	457.6	
907	Low voltage			6	1,500			4×6	10.437	265.10	2.718	69.03	7.218	183.33	148	199.0	3,108.4	17-414, EX-5590-7, BA-249 IJ, 13,255, 7MW, 4668, 7D8907
908	Low voltage	Lantern		6	250			4/6	2.625	66.67	2.625	66.27	4.406	111.91	21.5	26.5	413.9	17-442, EX-5550-5, BA-200/U, 13,257, 5MW, 4667, 7D8908
908C	Low voltage	Lantern		6	250	0.460	11.68	4/F	2.625	66.67	2.625	66.27	4.406	111.91	23.0	27.3	426.4	4702
910A	Alkaline		L-20	1.5	85	0.445	11.30	1×340					1.120	28.44	0.35	0.18	2.8	
910F	Low voltage		N	1.5	20	0.460	11.68	1×4					1.180	29.97	0.22	0.16	2.5	
910M			M-35	1.4				1×401					1.120	28.44	0.4	0.18	2.8	
911			6	1.5		2.625	66.67	1×6										6443
912	Low voltage	Ignition		7.5	1,500			5×6	7.843	199.21	4.968	126.18	7.218	183.33	34	29.3	457.7	BA-23
913	Low voltage	Ignition		9	1,500			6×6	7.812	198.42	5.281	134.14	7.250	184.15	180.0	273.0	4,264.3	
915	Low voltage	Ignition		6	250			4/F	2.625	66.67	2.625	66.67	3.843	97.61	216.0	290.0	4,529.8	BA-206 IJ
917	Low voltage	Miscellaneous		6	250			4/F	2.625	66.67	2.625	66.67	4.937	125.40	23	26.5	413.9	4677, 7D-8915
918	Low voltage			12	250			8/F	5.343	135.71	2.843	72.21	4.437	112.69	20	26.5	413.9	
920	Low voltage	Flasher		6	250			4/F	2.625	66.67	2.625	66.67	3.843	97.61	52	69.5	1,085.5	17-443, EX-5560-3, 13,274, 8MW, 470
922		Fence controller		12	600			24/E							23	27.3	426.4	
923		Flasher		12	500			8/F	5.343	135.71	2.843	72.21	4.156	105.56	133	199.0	3,108.4	5704
926				12	250			8/F	5.343	135.71	2.843	72.21	4.437	112.69	48	63.1	985.6	
927	Alkaline		L-80	1.5	300	1.343	34.11	1/L80					1.05	26.67	52	69.5	1,085.5	
930	Alkaline			6	1,300			4/L100	5.531	140.48	4.656	118.26	2.094	53.18	2.1	1.6	24.9	
															40.0	53.9	841.9	

49

TABLE 18 Characteristics of Transistor Batteries

NEDA	Designation	Type	ANSI designation	Voltage	Current, ma	Service, mah	Diameter in.	Diameter mm	Number of cells and sizes	Length in.	Length mm	Width in.	Width mm	Overall height in.	Overall height mm	Weight, oz	Volume cu in.	Volume cu cm	Other designations
1100	Instrument	Mercury	M40	1.4	100	1,000	0.620	15.70	1/M40					0.640	16.25	0.43	0.20	3.12	
1101	Instrument	Mercury	M70	1.4	250	3,600	0.640	16.25	1/M70					1.968	49.98	1.4	0.60	9.37	
1104	Transistor radio	Mercury	M20	1.4	2	350	0.610	15.49	1/M20					0.233	5.91	0.17	0.07	1.09	
1105	Transistor radio	Mercury	M30	1.4	50	500	0.620	15.70	1/M30					0.435	11.04	0.28	0.13	2.03	7D-9016
1106	Transistor radio	Mercury	M10	1.4	7	750	0.450	11.43	1/M10					0.130	3.30	0.04	0.02	0.31	
1107A	Light meter	Silver	S16	1.5	240	165	0.455	11.55	1/S16					0.220	5.58	0.09	0.03	0.46	
1108A	Transistor	Alkaline			60		0.563	14.30	1/1/2AA					1.021	25.93	0.40	0.24	3.75	
1200	Transistor	Mercury		2.7	100	1,000	0.656	16.66	2/M40					1.300	33.02	0.94	0.43	6.72	
1300	Transistor	Mercury		4.2	60	2,200	1.000	25.40	3/M60					1.968	49.98	3.15	1.60	24.99	
1303	Transistor radio	Zinc-carbon		4.5	300				6/D	2.812	71.42	1.750	44.45	8.343	211.91	25.0	41.0	640.42	6415
1305	Transistor	Mercury		4.2	50	500	0.656	16.66	3/M30					1.312	33.32	0.9	0.44	6.87	
1306	Transistor	Zinc-carbon		4.5	12		0.662	16.81	6/F12					1.965	49.91	0.9	0.65	10.15	
1306A	Transistor	Alkaline		4.5	70		0.662	16.81	3/125					1.956	49.68	1.17	0.67	10.47	7D-7806
1306M	Transistor	Mercury		4.2	100	1,000	0.656	16.66	3/M40					1.950	49.53	1.42	0.65	10.15	
1307	Snap type	Alkaline		4.5	70		0.662	16.81	3/125					1.923	48.84	1.25	0.67	10.47	
1308	Snap type	Alkaline		3	70		0.664	16.86	2/125					1.304	33.12	0.8	0.46	7.18	
1310A	Transistor	Alkaline		3	60		0.562	14.27	2/1/2AA					2.000	50.8	0.75	0.47	7.34	

TABLE 19 Characteristics of Radio Transistor and Transceiver Batteries

NEDA	Designation	Type	ANSI Designation	Voltage	Current, ma	Service, mah	Diameter in.	Diameter mm	Number of cells and sizes	Length in.	Length mm	Width in.	Width mm	Overall height in.	Overall height mm	Weight, oz	Volume cu in.	Volume cu cm
1400	Portable transistor	Zinc-carbon	......	6.75/7.00	80	......	......	......	16/F100	2.562	65.07	3.031	76.99	7.906	200.81	39.0	41.5	648.23
1403	Pocket transistor	Zinc-carbon	......	6	32	......	......	......	16/F22	1.406	35.71	4.109	104.36	4.437	112.69	3.5	4.58	71.53
1404	Transistor	Mercury	4M30	5.6	......	......	......	......	4/M30	......	......	......	......	1.752	44.50	1.2	0.61	9.53
1406A		Silver	......	6	10	165	0.656	16.62	4/S15	......	......	......	......	0.990	25.14	0.5	0.18	2.81
1500	TV tuner	Mercury	......	7	50	500	0.510	12.95	5/M30	......	......	......	......	2.192	55.67	1.5	0.73	11.40
1501	Transistor radio	Mercury	......	6.75/7.00	10	160	0.656	16.62	5/675	......	......	......	......	1.094	27.78	1.0	0.22	3.44
1600	Transistor radio	Zinc-carbon	6F24	9	9	......	0.500	12.70	6F/24	......	......	......	......	1.672	42.46	2.0	1.31	20.46
1600M	Radio	Mercury	......	8.4	8.4	750	1.000	25.40	6/822	......	......	......	......	1.937	49.20	1.6	1.52	23.74
1601	Portable transistor	Zinc-carbon	......	3, 6, 9	50	......	1.000	25.40	6/D	2.812	71.46	1.562	39.67	8.00	203.20	1.6	1.52	23.74
1602	Portable transistor	Zinc-carbon	6F50-2	9	150	......	......	......	6/148	1.406	35.71	1.344	34.13	2.656	67.46	4.5	5.0	78.10
1603		Zinc-carbon	6F100	9	15	......	......	......	6/F100	2.562	65.07	2.031	51.56	3.063	77.80	15.0	16.3	254.60
1604	Portable transistor	Zinc-carbon	6L10	9	30	......	......	......	6/F22	1.031	26.18	0.656	16.66	1.765	44.83	1.5	1.20	18.74
1604D	Transistor radio	Zinc-carbon	6F22	9	8	......	......	......	6/F22	1.031	26.18	0.656	16.66	1.765	44.83	1.5	1.20	18.74
1604M	Transistor radio	Mercury	6M25	8.4	30	575	......	......	6/635	1.031	26.18	0.660	16.76	1.640	41.65	2.0	1.5	23.43
1605	Transistor radio	Zinc-carbon	6F96	9	20	......	......	......	6/F90	1.812	46.02	1.812	46.02	2.343	59.51	7.0	7.70	120.27
1606A	Transistor radio	Mercury	7M16	9.8	10	200	0.556	14.12	7/S15	......	......	......	......	1.640	41.65	0.8	0.39	6.09
1608	Portable transistor	Zinc-carbon	......	9	30	......	......	......	6/D	2.281	57.93	1.250	28.57	7.968	202.38	19	22.7	354.57
1609		Zinc-carbon	......	9	150	......	......	......	6/D	2.812	61.11	1.750	44.45	8.343	211.91	25	41.0	640.42
1610	Transistor	Zinc-carbon	......	4.5	40	......	......	......	6/146	1.406	35.71	1.343	34.11	2.750	69.85	4.5	5.02	78.41
1611	Transistor	Zinc-carbon	6M22	9	......	......	0.750	19.05	6/F17	......	......	......	......	1.953	49.60	1.2	0.85	13.28
1611M	Transistor	Mercury	......	8.4	40	600	0.724	18.38	6/M26	......	......	......	......	2.000	50.80	1.65	0.82	12.80
1612	Portable transistor	Zinc-carbon	......	9	80	......	......	......	6/C	2.187	71.42	1.156	29.36	6.250	158.75	11.0	17.6	274.91
1613		Zinc-carbon	......	9	16	......	......	......	12/117	1.406	35.71	0.734	18.64	4.687	119.05	2.5	4.57	71.38
1614	Depth finder	Mercury	......	8.1/8.4	250	3,600	......	......	6/12	2.187	71.42	0.875	22.22	4.438	112.72	9.34	8.3	129.64
1810	Communication	Zinc-carbon	......	12	8	127	1.000	25.4	8/127	......	......	......	......	2.438	61.92	2.30	1.68	26.24
1810M	Transceiver	Mercury	......	12.6	50	750	1.000	25.4	9/822	......	......	......	......	2.406	61.11	3.6	2.18	34.05
1900	Portable transistor	Zinc-carbon	......	13.5	10	......	......	......	9/135	1.329	33.75	1.031	26.18	2.687	68.24	3.00	3.61	56.39
1901	Instruments	Mercury	......	12.5	50	500	0.662	16.81	9/640	......	......	......	......	3.984	101.19	4.2	1.37	21.40
2200M	Transceiver	Mercury	......	16.8	200	2,400	......	......	12/502AA	2.810	71.37	2.850	72.39	1.320	33.52	12.8	9.50	148.39

The zinc-carbon system has long been the common denominator for cells and batteries. Basic materials are in free supply and compared to the materials in other systems, such as silver, zinc, mercury, and regular alkaline-manganese cells, are less expensive. Manufacture and construction of zinc-carbon cells and batteries has been practiced for a long time. While other systems have certain advantages over zinc-carbon in that cells are of smaller sizes, can better stand cold weather, or leak less, the end user pays more for these advantages. It is difficult to fault the zinc-carbon system as an economical and efficient power source except in special cases, such as hearing aids, instruments to be used in cold weather, or instruments which require extra capacity.

Recharging Primary Batteries

It is possible to recharge a primary battery,[6] but only for a limited number of cycles and under controlled conditions. To be economically practical, battery recharging should be done on a large-scale basis. A zinc-carbon battery before recharging must have a working voltage not less than 1 v. The battery should be charged very soon after removal from service. The ampere-hours of recharge should be 120 to 180% of the ampere-hour discharge, and the recharge should take place over a period of 12 to 16 hr. In addition, the battery should be put into service as soon as it has been recharged, since such cells have a very poor shelf life.

Magnesium Batteries

Magnesium dry cells have been patterned[7] after the zinc cell. The cell contains a magnesium alloy (AZ31A alloy: Mg, 3Al, 1Zn, 0.2Mn, 0.15Ca) anode, MnO_2 + $BaCrO_4$ + acetylene black cathode, magnesium bromide electrolyte, and a paper separator. The cell has high capacity when the load resistance is designed for the voltage level of the cell. In series batteries, the optimum number of cells is 15 to 30% less than the number of Zn cells for the desired voltage. Cells show storage life of two years, retaining more than 85% of their initial capacity.

The electrolyte contains $MgBr_2$, 190 to 250 g/l, with lithium chromate (0.20 g/l) as an inhibitor. The carbon is surrounded by MnO_2 with acetylene black modified by 3% $BaCrO_4$. Kraft-paper separators are employed. Anode efficiency is about 60%, compared to 90 to 95% for zinc. Open-circuit voltage is 1.8 v.

Magnesium cells have attributes which are not shared by the zinc-carbon system. Basically these are longer life and storage and the ability to withstand humidity and high heat. One can recognize the benefit to the military in operations around the world, particularly in southeast Asia where batteries are stored under all kinds of severe conditions. Magnesium batteries are relatively more expensive than other systems because of materials costs, and high production quantities cannot reduce the cost compared to the zinc-carbon system from the present ratio of three to five times their zinc-carbon counterpart.

Methods of testing dry cells were first recommended in 1912.[8] This was followed by Bureau of Standards Characteristics and Testing Methods of Dry Cells in 1919, 1923, 1927, American Standards in 1937, and the NBS in 1941, 1947, 1954, and 1959.

REFERENCES

1. N. C. Cahoon, *Trans. Electrochem. Soc.*, **92**: 159–172 (1947).
2. L. C. Copeland and F. S. Griffith, *Trans. Electrochem. Soc.*, **89**: 495–507 (1946).
3. N. C. Cahoon, *Trans. Electrochem. Soc.*, **92**: 168 (1947).
4. W. E. Ramsey, *J. Franklin Inst.*, **225**: 401–409 (1938).
5. Milton E. Wilke, *Trans. Electrochem. Soc.*, **90**: 433–440 (1946).
6. Letter Circular LC-965, U.S. Dept. of Commerce, NBS.
7. R. C. Kirk, P. F. George, and A. B. Fry, *J. Electrochem. Soc.*, **99**: 323–327 (1952).
8. American Standard Specification for Dry Cells and Batteries, ASA, sponsored by NBS, ANSI C18.1-1965.

chapter 5
The Zinc-Alkali-Manganese Dioxide Primary Batteries

Alkaline Cells

The electrochemical system of both primary and rechargeable alkaline cells comprises a zinc anode of large surface area, a manganese dioxide cathode of high density, and a potassium hydroxide electrolyte.

Up to 1955, alkaline dry cells differing in details of construction from the dry cells had not proven commercial. The alkaline cell system, however, when coupled with a zinc anode of high surface area, shows high depolarizing efficiency. On heavy or continuous drains the alkaline cell shows an advantage over conventional cells on a performance per unit of cost.

Construction of the alkaline cell is shown in Fig. 18. There is a high-density manganese dioxide depolarizer with a steel can current collector and a zinc anode of high surface area in contact with the electrolyte of potassium hydroxide. These cells have low internal resistance and impedance and high service capacity. They are hermetically sealed.

Nominal voltage of an alkaline–manganese dioxide primary cell is 1.5 v in standard N, AA, C, 1/2D, D, and G cell sizes. The closed-circuit voltage of an alkaline primary battery falls gradually as the battery is discharged (Fig. 19).

The service hours delivered by alkaline-manganese primary batteries are greater as the end-point voltage is lower. Service capacity remains

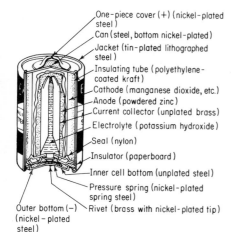

Fig. 18 Cutaway of alkaline cell (primary type). (*Eveready, Union Carbide Corporation.*)

relatively constant as the discharge schedule is varied. Capacity does not vary as much with current drain as it does in the zinc-carbon battery.

The alkaline system operates with high efficiency under continuous or heavy-duty, high-drain conditions where the zinc-carbon cell is unsatisfactory. Under certain conditions, alkaline cells will provide as much as

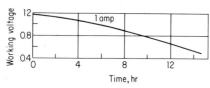

Fig. 19 Voltage discharge characteristic of alkaline-manganese primary battery D cell discharged 24 hr/day. (*Eveready, Union Carbide Corporation.*)

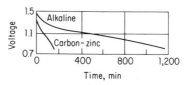

Fig. 20 Comparison of discharge characteristics of alkaline-manganese and zinc-carbon D-size cells for 500 ma continuous drain at 70°F. (*Eveready, Union Carbide Corporation.*)

10 times the service of standard zinc-carbon cells. Discharge characteristics of the two battery types are compared in Fig. 20.

Alkaline-manganese cells may not show an advantage at light drains and/or under intermittent-duty conditions. With intermittent use below 300 ma for the D-size cell, alkaline cells will lose their economic advantage over zinc-carbon batteries.

Alkaline cells are employed for camera cranking, radio-controlled model planes and boats, electronic flash, model boats and automobiles, and radios and tape recorders.

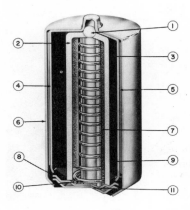

Fig. 21 Cross section of a typical cylindrical zinc–manganese dioxide dry cell. (1) Glass insulator. (2) Electrolyte absorbent. (3) Manganese dioxide cathode and depolarizer. (4) Inner gold-plated can. (5) Adapter sleeve. (6) Outer nickel-plated can. (7) Zinc anode. (8) Molded grommet. (9) Barrier. (10) Spring contact. (11) Double top assembly. (*Radio Corporation of America.*)

The RCA version is shown in Fig. 21, and the Burgess version is shown in Fig. 22. As shown in Fig. 22, a typical alkaline-manganese cell uses a cylindrical depolarizer on contact with the cell container of (usually nickel-plated) steel. Because of the passivity of steel in alkaline electrolytes, there is no chemical reaction between the MnO_2 depolarizer

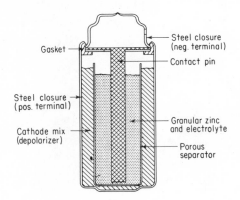

Fig. 22 Cross section of alkaline-manganese cell. Through external jacketing and terminal connections, cells can be made to appear conventionally polarized (center button +). (*Burgess Battery Company.*)

and the steel, permitting the latter to be a current collector as well as a strong cell container. The depolarizer surrounds a cylindrical granular zinc anode, the two electrochemical components being separated by porous sheet materials. When the zinc is amalgamated to discourage

TABLE 20   Alkaline Zinc–Manganese Dioxide Cells

Eveready	NEDA	Voltage	Current, ma	Service, mah	Number of cells and sizes	Average flash current, amp	Internal impedance, ohm	Diameter in.	Diameter mm	Length in.	Length mm	Width in.	Width mm	Height in.	Height mm	Weight, oz	Volume, cu in.	ANSI	Zinc-carbon equivalent	Nickel-cadmium equivalent
EP × 825		1.5		300	1/825			0.905	22.98					0.228	5.79	0.026	0.106			
								0.455	11.56					0.570	14.48	0.016	0.09	L10		
E90	910A	1.5	85		1/N	2.5	0.6	0.470	11.93					1.130	28.70	0.35	0.18	L20	N	
								0.625	15.88					0.660	16.76	0.40	0.20	L25		
E89	1108A	1.5	60		1/1/2AA	3	0.5	0.563	14.30					1.087	27.60	0.4	0.24			
E92	24A	1.5	100		1/AAA	7.5	0.2	0.410	10.41					1.745	44.32	0.4	0.16	L30	AAA	K40
E91		1.5	150		1/AA			0.562	14.27					1.937	49.19	0.75	0.48	L40	AA	K70
E93		1.5	480		1/C	8.3	0.18	1.031	26.29					1.969	50.01	2.3	1.64	L70	C	K70
								1.030	26.16					1.937	49.19	2.2	1.25	L70	C	K80
E94		1.5	300		1/1/2D			1.344	34.13					1.203	30.55	2.1	1.6	L80	1/2D	K90
E95		1.5	650		1/D	10.0	0.15	1.344	34.13					2.406	61.11	4.5	3.17	L90	D	K100
		1.5						1.344	34.14					4.360	110.74	8.0	6.19	L100	G	4K100
532	1308	3.0	70		2/L25			0.664	16.86					1.670	42.42	0.8	0.46	2L25		
529	1310	3.0	60		2/1/2AA			0.562	14.27					2.0	50.8	0.75	0.47	2/511		
523	1307	4.5	70		3/L25			0.662	16.81					1.965	49.91	1.17	0.67	3L25		
531		4.5	70		3/L25			0.662	16.81					2.290	58.16	1.25	0.67	3L25		
520		6.0	1,300	4,000	4/L100	Rechargeable				5.531	140.43	4.656	118.26	2.656	67.46	40.0	53.9	4L100	4G	4K100
596		1.5	1,000	4,000	1/460	Rechargeable								3.730	94.74	7.5	5.5			
563		4.5	625	2,500	3/L90	Rechargeable		1.359	35.43					6.968	176.98	15.0	10.1	4L90		
560		7.5	625	2,500	5/L90	Rechargeable		1.359	35.43					7.156	181.76	25.0	29.0	5L90		
564		13.5	1,250	5,000	9/L100	Rechargeable				2.656	67.46	1.531	38.88	5.875	149.22	88.0	137.0	9L100		
561		15.0	1,250	5,000	10/L100	Rechargeable				8.312	211.12	2.812	71.42	5.875	149.22	96.0	137.0	10L100		

gas generation on open-circuit stand, copper-coated steel or brass current collectors are generally used, and the cell may be hermetically sealed. However, some manufacturers provide rupturable closures or relief valves to avoid violent gas relief in the event of a faulty or misused cell.

Recommended sizes as well as equivalents of the cylindrical alkaline–manganese dioxide cells are given in Table 20; commercial units, with manufacturing tolerances, capacities, and equivalents to the ANSI designations are also given.

chapter 6

Air-depolarized Cells

A transition form bridging the alkaline dry cells and in a measure the fuel cell is the air cell.

Porous and Adsorptive Carbon Electrodes

The air-depolarized cell comprises a caustic alkali electrolyte, an anode of amalgamated zinc, and a carbon cathode capable of utilizing atmospheric oxygen. The operating voltage is of the order of 1.1 to 1.2 v for ordinary loads, and the open-circuit voltage of a fresh cell 1.4 to 1.5 v. The cathode is of nongraphitic carbon with powdered charcoal, manufactured so as to be sufficiently porous and gas-permeable to "breathe air" and yet close-grained enough to resist electrolyte penetration. Heise and Schumacher[1] state that the carbon has an apparent density of 0.6 to 0.8, a porosity of 50 to 65%, a resistivity of 0.0075 to 0.015 ohm/cu cm and an air permeability of 100 to 1,000 cu cm/min/cu cm of carbon. Oxide complexes at the electrode interface are considered to be in the depolarizer. The electrode is waterproofed with paraffin. The electrolyte is usually a 20% NaOH solution for low-temperature service.

60 Batteries and Energy Systems

Lime is added to react slowly with the sodium zincate formed, with the regeneration of caustic.[2]

Figure 23 shows a two-cell unit made for operation of 2-v radio tubes. These batteries are good for about a year of service when operated 3 hr/day. The voltage characteristics of the air cell are given in Fig. 24. A Union Carbide cell is shown in Fig. 25.

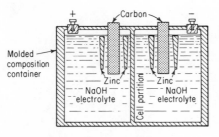 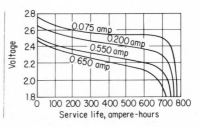

Fig. 23 Air cell—two-cell battery.

Fig. 24 Voltage characteristics of an air cell.

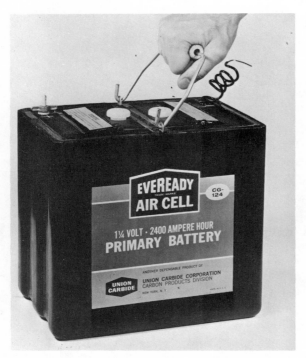

Fig. 25 Battery type CG-124, 1.25 v, 2400 ah. (*Eveready, Union Carbide Corporation.*)

The cell is shipped dry, with the caustic in cast form around the anode and cathode and with lime in dry form at the bottom of the container, and "activated" by addition of water.

Competitive Types and Application

Primary batteries of the adsorbent-carbon-depolarizing type are also built for semaphore, highway flashing systems, lighthouses, railway signals, and similar work. A typical one is the Le Carbone cell, the construction of which is shown in Fig. 26 and the characteristics in Fig. 27.

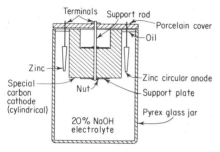

Fig. 26 Le Carbone caustic cell.

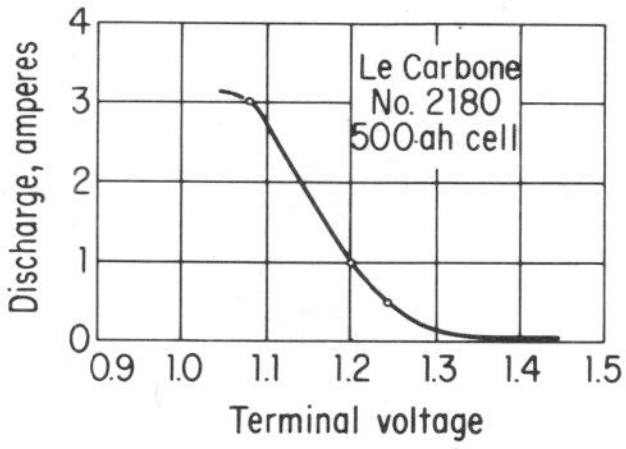

Fig. 27 Current-voltage relations of Le Carbone cell.

A cell whose jar is approximately 10 in. high and 7 in. diameter is rated at 500 ah, with initial closed-circuit voltage of over 1 v, at rates of discharge not greater than 3 amp. The carbon has a capacity of 2,000 to 2,500 ah, while the zinc, weighing about 825 g, lasts 500 ah, after which it is renewed. The cell holds about 4 l of a 20% NaOH electrolyte, which is renewed every 500 ah. The open-circuit voltage of the cell is 1.4 to 1.5 v.

Table 21 gives the characteristics of the line of Le Carbone cells. It is interesting to note that for convenience and utility without service, connection, or metering from a central station or power source, the customer is willing to pay on the order of $9 to $31 per kwh over the life of the battery. With renewal of the zinc, the cost is on the order of $10 per kwh over the life of the battery.

The zinc-air primary battery of Leesona Moos, Yardney, and Eagle-Picher may be mechanically recharged by replacing the plates. The anode is a porous zinc structure with a double separator paper attached to the electrode. The electrolyte is prevented from leaking by an O-ring seal to the plastic top on the anode. The anodes are soaked in a 36% solution of potassium hydroxide, dried, and the water removed leaving solid hydroxide. Activation is by water and insertion of the anodes into the chambers.

TABLE 21 Le Carbone Air Cells

Type of cell	Dimensions, in.	Weight, lb	Capacity, ah	Voltage	Continuous discharge, amp	Intermittent discharge, amp	wh	Price	Price per kwh	Dollars per lb
AN 110	7 × 7 × 8½	20	1,000	1.2	0.40	3	1,200	11.10	9.25	$0.56
AN 210		35	1,000	2.4	0.150	0.57	2,400	21.60	9.00	0.62
500	6⅜ × 6⅜ × 8	15	500	1.2	0.4	....	600	11.05	18.50	0.74
250	4¼ × 4¼ × 8	6.5	250	1.2	0.2	....	300	5.65	18.80	0.87
135	3¼ × 3¼ × 7½	3.25	135	1.2	0.15	....	162	3.35	20.68	
Instrument 250	4¼ × 4¼ × 8	6.5	250	1.2	0.20	....	300	9.50	31.67	1.46
Instrument 135	3¼ × 3¼ × 7½	3.25	135	1.2	0.08	....	162	5.05	31.17	1.55
ANRD	8 5/16 × 8 5/16 × 14¼	(dry) 41 (activated) 51	3,200	1.25	1.0					
2 ANS		(dry) 23.5 (activated) 30.5	1,200	2.5	1.0 (70°F) 0.25 (0°F)					

A 24-v, 25-ah system built by Leesona Moos for the military has 22 cells connected in series, each with a porous separator. This structure provided support and the spacing for air access.

Rate of discharge of a zinc-air cell is dependent upon the rate at which air can be admitted. For high-discharge applications, blowers are provided. The battery by Leesona Moos for the AN/PRC 41 and 47 radio transceivers weighs 11 lb and provides 60 wh/lb.

Another variation is the zinc-oxygen system under development at Leesona Moos. This unit delivers 150 wh/lb, at a 50- to 400-hr rate. Cryogenic oxygen will afford energy densities of 175 wh/lb.

These portable zinc-air batteries are being used in man-pack transceivers, night vision devices, and space satellite communications.

A new design is the rechargeable alkaline metal-air battery developed at General Telephone and Electronics Labs. The unit has a porous iron anode, with an air cathode, a permeable hydrophobic structure supported on a nickel grid containing platinum black. The electrolyte is 20 to 45% potassium hydroxide.

Iron-air cells with 5 and 20 ah have been fabricated and operated for 200 charge-discharge cycles. The discharge was 65%. The power densities are 60 to 70 wh/lb on continuous cycling.

The cell has two voltage plateaus. The first is 0.96 to 0.80 v, the second is between 0.70 and 0.60 v. The first plateau represents 65% of the anode capacity in ampere-hours.

Under development at General Electric is a magnesium-air primary battery that uses salt water as the electrolyte. The "Mag-Air" battery has magnesium as the anode. The porous cathode promotes the action of the oxygen in the air with the water in the electrolyte and the magnesium electrode to produce current. Magnesium hydroxide is a by-product. The battery has an energy density of about 50 wh/lb and consists of 23 cells.

One set of magnesium anodes supplies 24 v for a field radio for up to 12 hr before replacement. To reactivate, the anodes are removed, magnesium hydroxide sediment is emptied out, and fresh magnesium, water, and salt are added.

The batteries have been tested for 30 recharging cycles, or 400 hr, without requiring replacement of the air cathodes.

REFERENCES

1. G. W. Heise and E. A. Schumacher, *Trans. Electrochem. Soc.*, **62:** 383 (1932).
2. U.S. Patents 1,835,867 (Dec. 8, 1931); 1,864,652 (June 28, 1932).

chapter 7

Fuel Cells

Most of the electrical power in the world is generated by heat engines utilizing the heat from the combustion of fossil fuels, subject to Carnot cycle limitation.

Couples are not limited by the heat cycle and will convert at high efficiencies. When the solid active electrode materials are replaced by reducible and oxidizable gases continuously fed from outside sources, the system becomes a fuel cell.

The approximate thermal conversion efficiency of a fuel cell is 60 to 85%, of central steam power stations 35 to 40%, of automobile engines 17 to 23%, of gas turbines 30%, of outboard motor engines 12%. Figure 28 shows the theoretical energy densities of electrochemical couples.

Types and Preparation

For over a century it has been realized that electricity can be produced directly from the oxidation of carbon or carbonaceous gas in galvanic cells operated with fused salt electrolytes. Thermally stable materials and battery components for long life at temperatures of 500 to 750°C have proven elusive.

A typical system is the hydrogen-oxygen couple, with the highest theoretical output (1,620 w/lb of active reactants). No active gases or acids are utilized. The Union Carbide fuel cell operates at temperatures of 70 to 150°F at pressures from atmospheric to a few atmospheres.

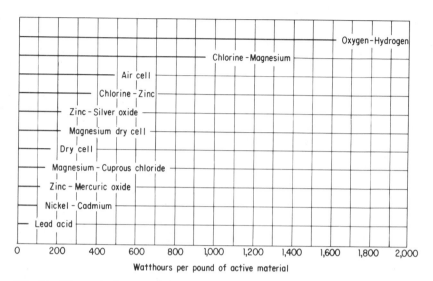

Fig. 28 Theoretical energy density of electrochemical couples. (*Union Carbide Consumer Products Company.*)

Basic forms are illustrated in Figs. 29 and 30. A flat-plate multicell battery is shown in Fig. 31.

Operation is continuous as long as fuel and oxygen are supplied, and in most instances the fuel supply settings remain unchanged after operation is started. Hydrogen and oxygen diffuse through the carbon electrodes to the electrolyte interface, where they remain as adsorbed gas phases until the electrochemical reactions take place.

The by-product of the hydrogen-oxygen fuel cell is water, formed at the carbon anode in amounts corresponding to the combined weight of oxygen and hydrogen consumed. The open-circuit voltage reaches a maximum of about 1.12 v instead of the theoretically expected potential of 1.20 v for the reactions

Cell reaction	$O_2 + 2H_2 \rightleftharpoons 2H_2O$	(1)
Cathode reaction	$O_2 + 2H_2O + 4e \rightleftharpoons 4OH^-$	(2)
Anode reaction	$2H_2 + 4OH^- \rightleftharpoons 4H_2O + 4e$	(3)

At active carbon cathodes, the cathodic reduction product of oxygen is peroxyl ion (HO_2^-)

$$O_2 + H_2O + 2e \rightleftharpoons HO_2^- + OH^- \tag{4}$$

which undergoes decomposition at the electrode surface to form hydroxyl ions, releasing oxygen for further electrochemical reaction

$$HO_2^- \rightleftharpoons \tfrac{1}{2}O_2 + OH^- \tag{5}$$

This gives rise to a four-electron process, and is reflected in hydrogen and oxygen consumption in operating cells corresponding to the requirements of reaction (1). Electrochemical efficiencies are 95% or better.

When the cell is operated at 140 to 150°F, the water is removable by circulation of an excess of hydrogen through the anode gas compartment and subsequent condensation of the vapor, the dry gas being returned to the cell. Some water vapor is found in the cathode system when the battery is operated at the higher temperatures, and this also may be removed by condensation. A schematic arrangement of battery and components is shown in Fig. 32.

Pressurized batteries operating at 5 to 10 atm with oxygen provide optimum power output. Maximum power density is approximately 1,900 w/cu ft of battery volume, but is obtained at the expense of excessive temperature rise and a sharp drop in fuel efficiency from normal operating conditions.

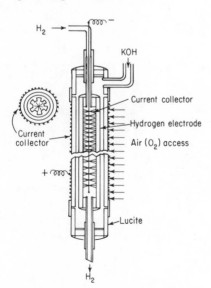

Fig. 29 Concentric tube construction. (*Union Carbide Consumer Products Company.*)

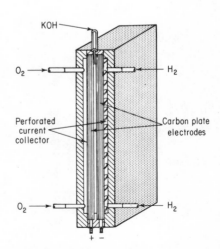

Fig. 30 Plate electrode construction. (*Union Carbide Consumer Products Company.*)

Fuel Cells 67

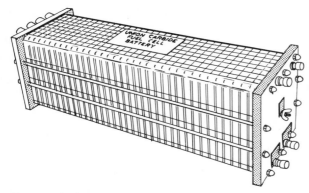

Fig. 31 Union Carbide plate-type fuel-cell battery.

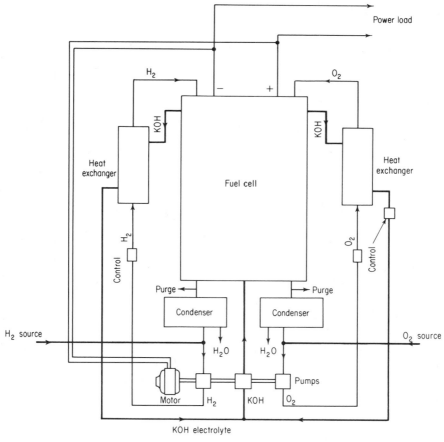

Fig. 32 Schematic diagram—typical H_2-O_2 fuel-cell battery. (*Union Carbide Consumer Products Company.*)

Figure 33 shows that a pressurized cell at 140°F will operate at 100 asf at about 0.10 v higher potential than a similar unit used at atmospheric pressure and 70°F. Thus a cell voltage of 0.92 v corresponds to 40 asf when oxygen is at atmospheric pressure and cell discharge proceeds at

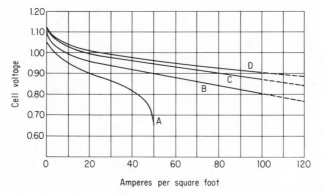

Fig. 33 Indicated performance of H_2-O_2 cells (operating voltage vs. current density). (a) 140°F; air at atmospheric pressure. (b) 70°F; O_2 at atmospheric pressure. (c) 140°F; O_2 at atmospheric pressure. (d) 140°F; O_2 at 5 atm pressure. (*Union Carbide Corp., Consumer Products Division.*)

70°F; the same potential is observed when the fuel cell operates at about 90 asf at the higher temperature with pressurized oxygen.

High rates of current withdrawal are not compatible with long life, and high current densities are suggested only where special service conditions dictate relatively short use periods and minimum weight and size. For long periods of operation, current density should not exceed 25 to 35 asf of active anode or cathode surface.

Table 22 illustrates variations in battery size and weight with operating conditions. Weight ranges from approximately 110 lb/kw for heavy operation to about 305 lb/kw where service requirements can be met with a higher load and cathodic depolarization with air instead of oxygen. The weights include auxiliary equipment such as gas blowers, water-vapor condensers, and piping but are exclusive of fuel and fuel storage containers.

Gas per kilowatt hour of use is shown in Table 23.

Fuels

Hydrogen is commonly stored as compressed gas in cylinders, or it may be produced from hydrides or reactive metals. No purification is required. Other fuels may be petroleum fraction, alcohols, ammonia, or simple

TABLE 22 Influence of Battery Operating Conditions on Weight and Volume

Characteristic	Battery rating kw	Air-hydrogen (1 atm, 70°F)	Oxygen-hydrogen (1 atm, 70°F)	Oxygen-hydrogen (1 atm, 140°F)	Oxygen-hydrogen (5 atm, 140°F)
Current density, asf	1	25	25	50	100
Terminal voltage per cell	1	0.88	0.94	0.94	0.90
Battery weight, lb*	1	305	290	166	110
Battery volume, cu ft*	1	4.4	4.1	2.4	1.4
Current density, asf	10	25	25	50	100
Terminal voltage per cell	10	0.88	0.94	0.94	0.90
Battery weight, lb*	10	2750	2600	1495	990
Battery volume, cu ft*	10	39	36	19	11

* Includes terminals, gas connections, and accessories but does not include fuels or fuel storage system.

organics. Oxygen can be taken directly from the air or supplied as compressed gas in cylinders. Carbonates are removed periodically from the electrolyte.

In the Allis-Chalmers version, structure is as shown in the expanded view in Fig. 34. There is an electrode support or current-carrying plate, made of corrosion-protected magnesium. This positions the electrode, serves as a manifold to distribute gas to the electrode area, and conducts current externally.

The anode is a sintered nickel plaque, catalyzed with a mixture of plati-

TABLE 23 Theoretical Reactant Requirement per kwh —Hydrogen-Oxygen Cell

Cell voltage (terminal)	Conversion* efficiency, %	Oxygen lb	Oxygen cu ft (NTP)	Hydrogen lb	Hydrogen cu ft (NTP)	Total gas lb	Total gas cu ft (NTP)
1.10	91.7	0.5988	7.250	0.0754	14.552	0.6742	21.802
1.00	83.3	0.6580	7.962	0.0829	16.000	0.7409	23.962
0.90	75.0	0.7310	8.845	0.0921	17.775	0.8231	26.620
0.80	66.7	0.8225	9.952	0.1036	19.995	0.9261	29.947
0.70	58.3	0.9400	11.374	0.1184	22.851	1.0584	34.225
0.60	50.0	1.0967	13.260	0.1382	26.673	1.2349	39.933

* 1.20 v is theoretical value for the $2H_2 + O_2 \rightleftharpoons 2H_2O$ reaction in 12 M KOH at 20°C.

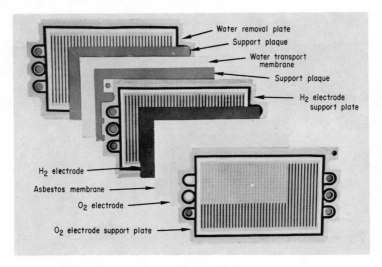

Fig. 34 Cell construction. (*Allis-Chalmers.*)

num and palladium. The cathode or oxygen electrode is powdered silver, supported in plastic. The capillary matrix filling the space between electrodes is asbestos saturated with 35% potassium hydroxide.

In operation, water is produced and must be continually removed. If too much water be removed, the cell will dry up. The components for water removal are water-transport membranes saturated with 45% potassium hydroxide, two porous support plaques, and a plastic water-removal plate similar to the support plate. Two end plates and tie bolts complete a single cell.

Figure 35 is an assembly drawing of a four-cell module.

The capillary potential of the electrolyte matrix—the differential pressure required to force liquid from its largest pore—is in excess of 100 psi, producing separation of gases and transport of water and hydroxyl ions between the electrodes. The matrix is compressed between the electrodes, which are supported by the oxygen and hydrogen current-carrying plates. Figures 36 and 37 show the simplified cell mechanisms.

The capillary potential of the electrode is about 6 psi less than the electrolyte matrix, and assures that flooding of the electrolyte will not occur by migration.

There is a stable electrolyte front produced with the capillary matrix. The matrix is filled with electrolyte during assembly, and when compressed forces electrolyte into the electrodes, maintaining the electrolyte front within the electrode. The volume in the electrode provides a reservoir for water and electrolyte, allowing operation of the cell through a range of 33 to 38% potassium hydroxide. Figure 37 shows the electrolyte front.

As water is produced, the effect on cell performance is shown in Fig. 38.

Control of water utilizes the vapor pressure characteristics of a KOH solution. At constant temperature, vapor pressure is directly proportional to electrolyte concentration. The static moisture-removal method

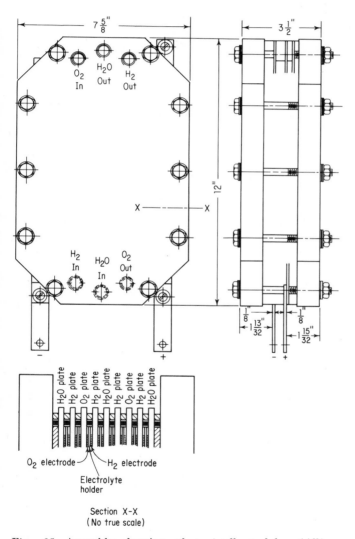

Fig. 35 Assembly drawing of a 4-cell module. (*Allis-Chalmers.*)

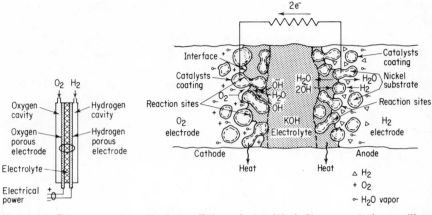

Fig. 36 Diagram of H_2-O_2 capillary fuel cell. Encircled portion is shown enlarged in Fig. 37. (*Allis-Chalmers.*)

Fig. 37 Enlarged simplified diagram of the capillary fuel-cell reaction mechanisms. (*Allis-Chalmers.*)

takes the water from the cell as a vapor while the recirculating electrolyte removes the water in the liquid form.

A membrane is positioned in contact with the hydrogen cavity, saturated with potassium hydroxide at a higher concentration than in the electrolyte. This transport membrane is backed up with porous support plaques which give mechanical strength and absorb volume changes owing to varying potassium hydroxide concentrations. The

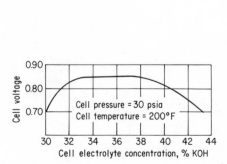

Fig. 38 Fuel-cell performance at various electrolyte concentrations. (*Allis-Chalmers.*)

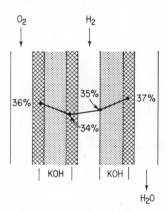

Fig. 39 Diagram of KOH gradients in fuel cell with static vapor pressure control. (*Allis-Chalmers.*)

membrane is filled with potassium hydroxide so that the loss of hydrogen by diffusion is negligible. Water diffuses through the hydrogen in an effort to reach equilibrium. A schematic of this gradient is shown in Fig. 39.

In static systems, a cavity adjacent to the water-transport membrane is at a pressure corresponding to the vapor pressure of the KOH solution, shown in Fig. 40. By lowering the cavity pressure, water is evaporated from the water-transport membrane and the concentration increases.

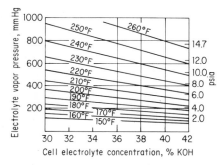

Fig. 40 Vapor pressure of aqueous solution of KOH. (*Allis-Chalmers.*)

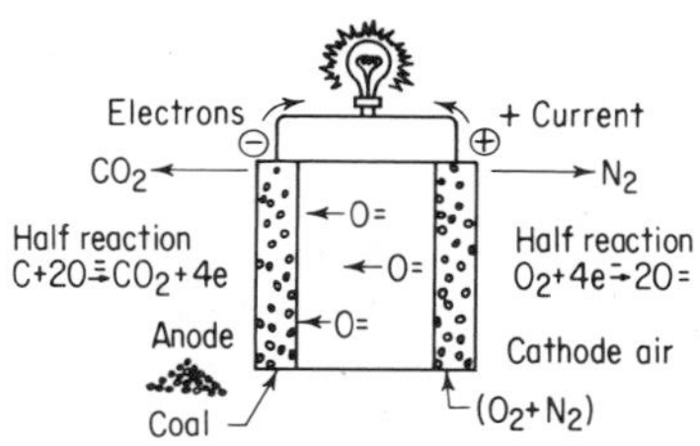

Fig. 41 Schematic carbon-oxygen direct fuel cell.

This increases the gradient across the hydrogen cavity and thus controls the rate of water removal from the cell electrolyte.

In the recirculating system, potassium hydroxide is circulated through the cavity. The potassium hydroxide concentration and, thus, water removal rate, is controlled by volume adjustment.

A large number of fuel-cell systems have been proposed and studied, involving alcohols, hydrazines, aldehydes, hydrocarbons of various sorts, and, earlier, solid fuels.

In the simplest fuel cell (Fig. 41) coal is supplied at the anode, where it interacts with oxide ions to form CO_2 and releases electrons to the external circuit. These electrons do work on the way to the cathode, where they are captured by oxygen from the air. The oxide ions thus formed complete the circuit by flowing through the electrolyte to the anode. With oxide ions carrying all the current, and with their production and consumption in balance, the electrolyte remains invariant.

Before 1900, "electricity direct from coal" was studied. Growth in the electrical industry brought home the advantages of electrical energy. Today, our best central stations operate at overall efficiencies near 40%. In 1896 the efficiency of a coal plant was 2.6%.[1]

In 1896,[1] Jacques, an American, made an attempt to take advantage of this opportunity by a fuel cell. The central carbon anode was surrounded by molten alkali in an iron pot that served as cathode. Air was directed to the walls of the cylindrical cathode by a distributor. Current densities over 100 asf at about 1 v could be maintained. As many as 100 of these cells were assembled to make the battery. This battery, rated at 1.5 kw, operated intermittently over a period near 6 months.

The possibility of fuel replenishment marks a distinction between fuel cells and conventional batteries. For short periods, conventional batteries are generally superior to fuel cells, which come into their own when electrical energy is required over long periods—periods long enough to necessitate continuous feed or replenishment of fuel.

In an article entitled "The Hydrocarbon Fuel Cell—Problems and Approaches," Liebhafsky and Cairns of the General Electric Company[2] state that

> Hydrocarbon fuel cells, and to a greater degree hydrocarbon fuel batteries, present formidable and interrelated heat, mass, and momentum transfer problems, all closely linked to chemical and electrochemical kinetics. These devices present to chemical engineering a challenge that will have to be met before fuel batteries can consume petroleum products in significant amounts.

They conclude

> We have seen that difficulties in meeting the reactivity requirements are the most serious obstacles in the road to a successful hydrocarbon fuel cell. These difficulties can be mitigated by increasing temperature, by proper choice of electrocatalyst and electrolyte, and by modifying the hydrocarbon. A fourth remedy, increasing the pressure, has not been discussed because it promises to be less rewarding than the others.
>
> In none of the five approaches discussed has the transition from the research stage (successful operation of a fuel cell in the laboratory) to the engineering development stage (combination of successful cells into successful fuel batteries) been completed. But it is not too early to begin considering the chemical engineering, outlined herein, that will have to be done before such batteries can be successful. The way in which this chemical engineering is done may determine how soon fuel batteries establish a new use for petroleum products.
>
> As a *tour de force*, one could even now operate a hydrocarbon fuel battery rated at a few kilowatts—after all, Jacques did the equivalent before 1900. Also, eventual success is likely for such batteries in at least some military and space applications, where cost is of secondary importance. In such applications, these batteries would be useful even if they failed to satisfy wholly the exacting requirements laid down above for complete success. Until the capital costs of hydrocarbon fuel batteries have been sharply reduced below what these costs would be today, it seems wise to postpone predictions about their commercial success.
>
> Markets for low-voltage d-c would be the first to be penetrated by a

TABLE 24 Typical Fuel-cell Systems: Hydrogen-Oxygen*

System	Electrodes	Electrolyte	Temperature, °C	Remarks
A. U.S.				
Union Carbide	C + catalysts	KOH	20–65	Low cost for catalysts.
Pratt-Whitney	Ni + catalysts	KOH	180–240	Project Apollo.
General Electric	Pt	ion membrane	20–80	Project Gemini; high cost for Pt catalysts.
Allis-Chalmers	Porous metals + catalysts	KOH	20–95	Electrolyte in porous material, special water control-removal system.
B. Foreign				
Raney nickel (Varta, USSR, Siemens, Brown-Boveri)	Raney nickel anode, usually silver or silver alloy cathode	KOH	20–90	Not as advanced as U.S. counterparts.
Quadrus (ASEA, Sweden)	Ni + catalysts	KOH	20–100	Submarine application anticipated; 200-kw unit under construction.
Tokyo Shibaura	Pd diffuser anode, Ni + catalysts cathode	KOH	30–100	Not as advanced as U.S. counterpart.

* Typical operating voltages for hydrogen-oxygen cells listed are 0.75–0.90 v. Hydrogen requirements are 0.087 to 0.105 lb/kwh and 0.70 to 0.84 lb/kwh oxygen depending on the operating voltage.

TABLE 25 Typical Fuel-cell Systems Other than Hydrogen-Oxygen

System	Electrodes	Electrolyte	Temperature, °C	Typical operating voltage	lb/kwh anode	lb/kwh cathode	State of development, remarks
A. Hydrazine-oxygen	Porous metals, carbon + catalysts	KOH	20–70	0.85	1.25*	0.75	Advanced: Union Carbide, Monsanto, Allis-Chalmers.
B. Sodium amalgam–oxygen	Steel anode; carbon + catalysts cathode	NaOH	20–70	1.50	1.22	0.43	Advanced: Kellogg, Yuasa, Japan Storage Battery.
C. Hydrocarbon-oxygen Pd diffuser anode	Pd diaphragm anode; Ni + catalysts cathode	NaOH	250	0.8	0.22†	0.80	Early: Pratt-Whitney.
Phosphoric acid	Pt or Pt alloys	Phosphoric acid	150–200	0.3	0.59†	2.1	Early: General Electric, Esso, others.
Fused carbonate	Porous metals + catalysts	Fused carbonates	500–650	0.7	0.25†	0.91	Intermediate: Texas Instruments, Institute of Gas Technology
Ionic conducting solid	Pt	Yttrium zirconate	800–1000	0.7	0.25†	0.91	Early: Westinghouse
D. Alcohol-oxygen	Pt or Pt alloys	Phosphoric acid	50–80°C	0.8	0.22†	0.80	Early: Esso, others.

* Hydrazine stored as a 60% by weight aqueous solution.
† Assuming butane or a higher saturated hydrocarbon.

successful central-station fuel battery, for these markets could use the output of such a battery without inversion. Fuel batteries for the home, with or without a conventional storage battery bank, would eliminate the cost of distributing the electricity they produce, but they would have to meet severe reliability and safety requirements.

Those interested in the commercial success of hydrocarbon fuel batteries will do well to watch their progress in military and space applications. The petroleum industry may some day have reason to be grateful for the fuel-cell research that military and space needs are currently justifying.

Vast research funds have been spent on fuel cells for the power plant of space vehicles, without limitation as to cost, because of the tremendous weight premium. They have been coupled with solar cells to provide 24-hr power, or with regenerative systems to operate radios, signal transmission, instrumentation, guidance, and control. Fuel-cell development is a favorite symposium topic for our chemical and engineering societies, space vehicle contractors, and defense agencies.

Yeager[3] reviewed fuel cells of the various fuel systems: only the hydrogen-oxygen, hydrazine-oxygen, and sodium amalgam–oxygen cells have advanced to the state where they are ready for application, as shown in Tables 24 and 25. One of the hydrogen-oxygen cells (General Electric ion membrane) has already found application as a nonpropulsive power source for orbital missions of NASA's Project Gemini. A second (Pratt-Whitney) was used in the circumlunar navigation mission of Project Apollo and a third (Allis-Chalmers) is now under study for later space missions. Researchers in Sweden have developed a hydrogen-oxygen cell which they plan to use for submarine propulsion as well as for other applications.

REFERENCES

1. W. W. Jacques, *Harper's*, **94**: 144 (Dec. 1896–May 1897).
2. H. A. Liebhafsky and E. J. Cairns, "Fuel Cells," pp. 50–56, AIChE, New York, 1963.
3. E. Yeager, *Chem. Eng. Progr.*, **44**(9): 92–96 (Sept. 1968).

chapter 8

Mercury Cells

During World War II there was a need for batteries to operate portable equipment requiring a high ratio of ampere-hour capacity to volume at higher current densities than were practical for the dry cell. Ability to withstand storage under tropical conditions was essential.

Types and Listing

The development, undertaken at the request of the U.S. Army Signal Corps, resulted in the R.M. or Ruben Cell,[1] comprising the system $Zn|Zn(OH)_{2(solid)}|KOH_{(aq)}|HgO_{(solid)}|Hg$ with a rated capacity of 200 mah for each 1.6-g unit of active material. The open-circuit voltage is nominally 1.34 v and the initial close-circuit voltage under normal loads varies from 1.24 to 1.31 v. The discharge curves on normal drains are substantially flat, with uniform voltage.[2]

There are two basic designs, one a roll anode and the other a pressed-powder anode.

A modification consists of a depolarizing mercuric oxide cathode, an anode of pure amalgamated zinc, and a concentrated aqueous electrolyte of potassium hydroxide saturated with zincate. The nominal voltage is 1.35 v with a depolarizer of 100% mercuric oxide and 1.4 v for a cell with a mixture of mercuric oxide and manganese dioxide. Utilization of active materials is 80 to 90%. Ninety percent efficiency is obtained at a current drain of 0.1 asi of depolarizer surface (0.2 ah/g of depolarizer).

The components are a pressed mercuric oxide cathode (in sleeve or pellet form) and pressed cylinders, or pellets, of powdered zinc with steel enclosures. These provide mechanical assemblies having dimensional stability.

Preparation

Designs are either flat or cylindrical pressed-powder electrodes. Depolarizing cathodes of mercuric oxide, to which a small percentage of graphite is added, are shaped as illustrated in Fig. 42 and either consolidated to the cell case (for flat electrode types) or pressed into the cases of the cylindrical types. Anodes are amalgamated zinc powder, pressed into flat or cylindrical shapes. A permeable barrier prevents migration of any solid particles in the cell. Insulating and sealing gaskets are of polyethylene or neoprene. Inner cell tops are plated to which zinc will form a zinc amalgam bond. Cell cases and outer tops of nickel-plated steel resist corrosion and provide passivity to internal cell materials. An outer, nickel-plated steel jacket encloses single-cell A battery types.

At moderately high pressures, the cell top is displaced upward against the external crimped edge of the outer jacket, tightening that portion of the seal and relieving the portion between the top and the inner steel cell case. Venting will then occur in the space between the internal cell container and the outer steel jacket. Cell electrolyte carried into this space will be retained by the absorbent ring. After venting, the cell stabilizes and reseats the top seal.

Characteristics of the system include: high capacity-to-volume ratio; flat discharge, sustained voltage under load; constant ampere-hour capacity; low internal impedance; the same capacity in either intermittent or continuous usage, and good shelf life. Cutaway views of flat and cylindrical cells are shown in Fig. 42.

Sizes and Applications

Batteries are available in voltages ranging from 1.35 to 97.2 v and in capacities ranging from 36 to 28,000 mah.

Mercury batteries have been used as secondary standards of voltage because of their better voltage maintenance and their ability to withstand mechanical and electrical abuses. As reference sources in regu-

80 Batteries and Energy Systems

lated power supplies, radiation detection meters, portable potentiometers, electronic computers, voltage recorders, and similar equipment, they exhibit several desirable features: voltage stability versus time, regulation within 0.5%, and for shorter periods, regulation of 0.1%; momentary

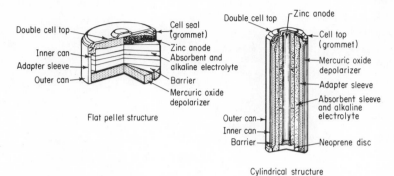

Fig. 42 Cross-section view flat- and cylindrical-type Eveready mercury cells. (*Eveready, Union Carbide Corporation.*)

short-circuits will cause no permanent damage; high drains without damage can be obtained; recovery to full open-circuit emf is rapid; the service capacity after 1 year of storage at 70°F is more than 90% of the capacity of the fresh battery.

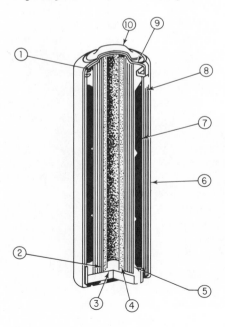

Fig. 43 Cross section of a typical cylindrical zinc–mercuric oxide dry cell: (1) sealing and insulating gasket, (2) electrolyte absorbent separator material, (3) insulator spacer, (4) powdered zinc anode, (5) microporous plastic, (6) outer steel case, (7) mercuric oxide–graphite cathode cylinders, (8) absorbent sleeve, (9) tin-plated steel inner top, (10) nickel-plated steel outer top. (*Radio Corporation of America.*)

In the RCA version of the mercury cell, red mercuric oxides with added graphite for increased conductivity are employed.

In the mercury cell shown in Fig. 43, the powdered-zinc negative electrode makes contact with a plated-steel cap which is insulated from

TABLE 26 Effect of Temperature on the Capacity of Mercury Dry Cells

Temperature, °F	Cell capacity, %
0	0
10	4.5
20	10
30	27
40	58
50	80
60	93
70	100
80	103
90	105
100	106
110	106
120	105
130	104
140	103

the outer steel case. This arrangement makes the cap, or the top terminal, negative with respect to the steel casing. Consequently, the polarity of the mercury cell is the reverse of that of the zinc-carbon cell.

The effect of temperature on capacity is shown in Table 26.

According to Burgess Battery Company, mercury cells are especially suited to provide maximum power output in minimum space.

A group of typical mercury cells, with ANSI, International Electrotechnical Commission, and NEDA designations, and listing maximum current (in milliamperes), service capacity (in milliampere-hours), cell or battery construction, dimensions, weight, and volume, have been brought together in Table 27. Note that a number of ANSI designations have not been completely accepted by other agencies or merchandisers. From M100 on, the ANSI designations are in a sense synthetic but they have found acceptance by engineers, designers, and electronic specialists as components in industrial electronic instruments. Note also the small sizes of cells for electric watches, with service lives on the order of a year, and small-size units for transistor application, instruments, walkie-talkies, depth finders, and transceivers. Their flat discharge curves make them an ideal source of power where voltage regulation is an important consideration. They provide the following desirable operation character-

TABLE 27 Mercury Dry Cells

Voltage	Application	ANSI designation	IEC designation	Other designations	NEDA	Current, ma	Service, mah	Number of cells and sizes	Diameter in.	Diameter mm	Length in.	Length mm	Width in.	Width mm	Height in.	Height mm	Weight, oz	Volume cu in.	Volume cu cm
1.35	Instruments	M5		313		5	40	1M5	0.310	7.87					0.140	3.56	0.03	0.01	0.2
1.40	Hearing aids, watch cells	M8		575		10	100	1M8	0.455	11.56					0.130	3.30	0.07	0.02	0.3
1.40	Instruments, watch cells	M10	MR-08	400	1106				0.455	11.56					0.135	3.43	0.04	0.02	0.3
1.40	Miscellaneous	M12		520		10	130	1M12	0.495	12.57					0.286	7.26	0.07	0.06	1.0
1.40	Instruments, watch cells	M15	MR-07	421, 675		10	180	1M15	0.455	11.56					0.210	5.33	0.09	0.03	0.5
1.40	Watch cells	M16							0.490	12.45					0.220	5.59	0.08	0.04	0.7
1.40	Watch cells	M17							0.600	15.24					0.150	3.81	0.08	0.04	0.7
1.35/1.40	Instruments	M20	MR-9	625		20	250	1M20	0.615	15.62					0.240	6.10	0.15	0.07	1.1
1.40		M22							0.592	15.04					0.305	7.75	0.17	0.08	1.3
1.40	Watch cells	M23							1.005	25.53					0.110	2.79	0.15	0.09	1.5
1.40	Transistor radio	M25	MR-01	450	1104	30	350	1M25	0.455	11.56					0.570	14.48	0.18	0.09	1.5
1.40		M26		660					0.685	17.40					0.305	7.75	0.26	0.11	1.8
1.40	Transistor radio	M30		640	1105	50	500	1M30	0.625	15.88					0.440	11.18	0.28	0.13	2.1
1.40		M31							0.888	22.56					0.229	5.82	0.29	0.14	2.3
1.40		M34							0.910	23.11					0.235	5.97	0.40	0.15	2.5
1.40	Instruments, radio	M35	MR-1	401		75	800	1M35	0.470	11.94					1.130	28.70	0.40	0.20	3.3
1.40		M36							0.406	10.31					1.750	44.45	0.40	0.23	3.8
1.40	Instruments	M40	MR-7	1		100	1,000		0.625	15.88					0.660	16.76	0.43	0.20	3.3
1.40	Instruments	M50							0.640	16.26					1.140	28.96	0.78	0.37	6.1
1.35		M55	MR-6	502		200	2,400	1M55	0.546	13.87					1.950	49.53	1.05	0.46	7.5
1.40		M60	MR-17						0.985	25.02					0.660	16.76	0.93	0.50	8.2
1.40		M62							1.195	30.35					0.510	12.95	0.57	0.50	8.3
1.40	Instruments	M70	MR-8	12	1101	250	3,600	1M70	0.640	16.26					1.955	49.53	1.40	0.63	10.3
1.40		M72	MR-19						1.225	31.12					0.660	16.76	1.60	0.78	12.8
1.40	Instruments	M100		42		1,000	14,000	1M100	1.281	32.54					2.390	60.71	5.85	3.08	50.5
2.80		2M25		2/450		30	350	2M25	0.500	12.70					1.25	31.75	0.42	0.19	2.97
2.80		2M35		2/401		75	800	2M35			0.985	25.02	0.510	12.95	1.23	31.24	0.88	0.50	7.81
2.80		2M40		2/1	1200	100	1,000	2M40	0.656	16.66					1.30	33.02	0.94	0.43	6.72
4.20		3M30		3/640		50	500	3M30	0.656	16.66					1.312	33.27	0.9	0.44	6.87
4.20		3M40		3/1	1306M	100	1,000	3M40	0.656	16.66					1.950	49.53	1.42	0.65	10.15
4.20		3M60		3/3	1300	60	2,200	3M60	1.000	25.40					1.969	50.01	3.15	1.60	24.99
5.60		4M30		4/640		50	500	4M30	0.656	16.66					1.752	44.50	1.20	0.61	9.53
5.60		4M40		4/1	1404	100	1,000	4M40	0.656	16.66					2.600	66.04	1.89	0.80	12.50
7.00		5M30		5/640	1500	50	500	5M30	0.656	16.66					2.192	55.68	1.50	0.73	11.40
7.00					1501	10	160	5M15	0.500	12.70					1.110	28.19	1.0	0.22	3.44
8.40	Transistor radio	6M26		6/660	1611M	40	600	6M26	0.724	18.39					2.000	50.80	1.65	0.82	12.81
8.40	Depth finder				1614	250	3,600	6M70			2.187	55.55	0.875	22.23	1.640	41.66	9.34	8.3	129.65
9.80				7/421	1604M	10	200	7S15	0.546	13.87					1.437	41.66	0.8	0.39	6.09
12.60	Transceiver				1810M	50	500	9/822	1.000	25.40					2.406	61.11	3.6	2.18	34.05
16.80	Transceiver				2000M	200	2,400	12/502AA			2.810	71.37	2.850	72.39	1.320	33.53	12.8	9.5	148.39

TABLE 28 Expected Service at Various Temperatures

Discharge temperature, °F	Approximate service expressed as percent of service at 70°F	
	Light drain,* %	Heavy drain,† %
113	100	100
70	100	100
40	95	7
28	6	2

* A light drain may be defined as a drain which will result in 100 hr or more service on a given cell.

† A heavy drain may be defined as a drain which will result in 30 hr or less on a given cell.

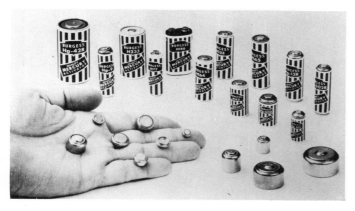

Fig. 44 Typical cell construction. (*Burgess Battery Company.*)

istics: (1) high ratio of energy to volume and weight, (2) long shelf life, (3) no need for recovery periods during discharge, and (4) relatively constant potential during discharge.

Table 28 shows the service that may be expected at various temperatures, with service at 70°F expressed as 100%.

Typical cell construction is shown in Fig. 44.

REFERENCES

1. Samuel Ruben, *Trans. Electrochem. Soc.*, **92**: 183 (1947).
2. Maurice Friedman and Charles E. McCauley, *Trans. Electrochem. Soc.*, **92**: 195 (1947).

chapter 9

The Silver Batteries

Types

Denison[1] described experimental zinc–silver oxide alkali cells with an electrolyte of 25% KOH in which electrolytically formed Ag_2O_2 served as one electrode and zinc as the other. The sensitivity of Ag_2O_2 to reduction by materials of construction in the cell was a severe disadvantage, but the discharge at a very high rate was an advantage.

A battery based on this system was patented in 1910 by Morrison,[2] and a patent was granted for a similar battery to André[3] in 1943. In 1902 Jungner[4] patented a battery having silver oxide positives and cadmium negatives. Later Jirsa[5] investigated iron.

The reaction of the system is

$$Ag_2O_2 + 2Zn + 2KOH \rightleftharpoons 2Ag + 2KHZnO_2$$

The milliampere-hour capacity is the same or a little greater than that of mercury batteries, but the silver oxide battery operates at 1.5 v while the mercury batteries operate at about 1.3 v. Types are available in 1.5-v cells in capacities of 60 to 165 mah, and 9 v at 165 mah. Current output ranges up to 10 ma.

Manufacture

The basic silver-zinc system in the charged state is silver peroxide as the positive material and zinc metal as the negative material. The zinc plate is grey, spongy, metallic zinc. The electron vehicle or electrolyte is a solution of potassium hydroxide. The active materials are separated from each other by wettable and semipermeable organic membranes such as cellophane or polypropylene with the addition of absorbents such as rayon or nylon. The couple provides an open-circuit (no load) voltage of approximately 1.82 v (single-cell voltage) upon initial immersion in the electrolyte; the voltage declines to lower levels depending upon time, temperature, and discharge rates. When completely discharged the silver peroxide is reduced to silver and the zinc is oxidized to potassium zincate. Figure 45 illustrates a typical discharge profile for this type of battery.

The electrolyte is potassium hydroxide in hearing-aid batteries. Sodium hydroxide is used in watch batteries for long-term reliability. Mixtures of silver oxide and manganese dioxide may be tailored to provide a flat discharge curve or increased service hours.

A range of silver batteries has been employed in ballistic missile programs.

Morse[6] states that the silver-zinc system supplying electrical power for today's ballistic missiles was not even considered practical at the close of World War II. Now this system, which exceeds all other battery systems in power-to-weight ratio, is considered among the most versatile and reliable. At a 100-sec rate of discharge, positive plate efficiencies exceeding 60% of theoretical are not uncommon, and at a 100-hr rate of discharge, capacities exceeding 90 wh/lb have been attained.

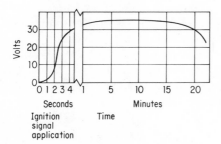

Fig. 45 Typical performance profile (80°F). (*Eagle-Picher Industries, Inc.*)

In the initial phases of missile and rocket development, secondary batteries were employed for electrical power sources, but as the primary silver-zinc battery has developed, it has been gradually replacing the secondary systems. The silver oxide–zinc system has been known as a feasible electrochemical couple since the mid-nineteenth century, but it continued to remain a laboratory curiosity until sometime in the 1920s or 1930s. Professor André of France was among those to receive early patents on the system as a secondary or storage battery. Some joint military-industry effort toward the development of a dry-charged system

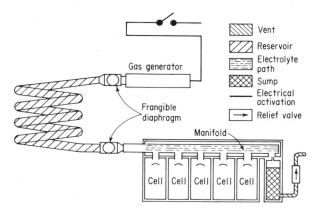

Fig. 46 Diagrammatic sketch of the system. (*Eagle-Picher Industries, Inc.*)

is reported to have been conducted during the early 1940s, but it was the latter part of the decade before the initial primary battery concept received consideration.

Although secondary batteries, including some silver-zinc types, had been used in early missile development, as soon as the primary reserve-type battery became available, inherent advantages dictated its use. The elimination of formation and charging procedures and the equipment associated was attractive. The cells were capable of high-rate discharge.

Silver batteries were employed in many of the early missiles and rockets under development by the Army. The first successful short- and intermediate-range ballistic missiles utilized power packs of these cells. These units became a part of the reliable hardware used to send the first American satellite into orbit in the early phase of our space age.

Following closely after the applications of manually activated primary silver-zinc batteries came efforts to provide automatic or remotely controlled activation. The principles include the transfer of electrolyte from a sealed reservoir to the cells of the battery as quickly as possible and in the simplest manner. Eagle-Picher developed a pressure-type system utilizing a copper tube electrolyte reservoir with a gas-generating cartridge.

In Fig. 46, the electrolyte is sealed into the reservoir with frangible diaphragms. When the gas generator receives its ignition pulse, the gas produced bursts the diaphragms at both ends of the electrolyte coil, and the electrolyte is forced from the tube into the battery manifold and on into the cells by the expanding gases.

Batteries activated with the metal tube reservoir system appeared to create a heavy system. One of the first efforts at meeting a specification celling for lightweight construction involved a requirement for a guided rocket. The weight limitation on this battery had already been set with a vacuum-activated, rubber bladder activation system, and it was necessary to make an equivalent design, weight-wise, with the coil system in order to meet requirements. Lightweight tubing and a foam-type potting compound in conjunction with a general weight-reduction effort on all internal components resulted in the development of a unit which meets these specifications. Voltage regulation has proceeded gradually upward.

The curves of Fig. 47 show the regulation available from a five-year-old design (at bottom) and a recent design. The GAP-4014 curves represent a regulation of about 11%, whereas the GAP-40XX represents better than 5%.

The early Redstone battery provided an output of 20 wh/lb. The next unit was the first remote-activated batteries on a ballistic missile. Its output dropped to 10 wh/lb.

Silver batteries provide guidance and hydraulic-control power systems for the Redstone Jupiter, Pershing, Atlas, and several others.

Comparison of the various silver systems may be seen in Table 29; other systems are included for comparison.

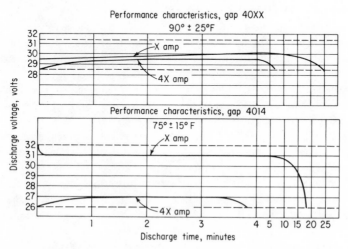

Fig. 47 Time discharge of voltage curves. (*Eagle-Picher Industries, Inc.*)

TABLE 29 Comparison of Various Silver Systems

Battery system	Open-circuit voltage	Working voltage	Operating temperature, *°F		Typical energy density†		Major application	Availability
			Normal	With heater	wh/lb	wh/cu in.		
Silver-zinc, primary	1.85	1.30–1.55	−40 to 130	−65 to 130	50–100	3.5–7.5	Where long, dry shelf life, maximum energy, and close voltage regulation are required. For extremely high continuous and pulse rates of discharge—life limited to 1 to 2 months.	Capacities 0.5 to 350 ah. High, medium, and low rate units. Capable of 1-min rates of discharge. Delivered dry-charged, custom designed to any ah size.
Silver-zinc, secondary	1.85	1.30–1.55	−40 to 110	−65 to 110	30–80	2.0–4.5	Where maximum energy density and voltage regulation are required throughout a cycle life period of 6 to 8 months.	Capacities 0.5 to 350 ah. High and low rate units delivered dry-charged, dry-uncharged, wet-charged, wet-uncharged-vented, sealed, custom designed to any ah size.
Silver-zinc, remote activated	1.85	1.20–1.55	32 to 120	−65 to 120	5–40 (As battery)	0.4–3.5 (As battery)	Long, dry shelf life. Unsurpassed in energy density and voltage regulation where activation by remote control is required. Rugged construction.	Any ah size. Over 150 battery sizes available.
Silver-cadmium, secondary	1.40	0.80–1.10	−40 to 110	−65 to 120	15–40	1.0–2.8	Where maximum energy density and voltage regulation are required throughout a cycle life period of 6 to 36 months.	Capacities 0.5 to 1.75 ah. High and low rate units. Delivered dry-charged, dry uncharged, wet-charged, wet-discharged-vented, sealed, custom designed to any ah size.
Nickel-cadmium, secondary	1.30	1.00–1.25	−65 to 120	−65 to 120	10–25	0.8–1.5	Where reliable cycling capability is required over a period measured in years.	Capacities 0.5 to 70 ah. Vented, sealed, customized battery packages.
Thermal, remote-activated	2.70 (Typical)	2.20–2.60 (Typical)	−65 to 200	Not required	0.5–4.0 (As battery)	0.04–0.3 (As battery)	Where unlimited shelf life and instantaneous activation are required at temperatures of −65 to 200°F.	Sizes from 5 to 500 v.
Water-activated CuCl	1.50–1.60	1.20–1.40	−50 to 200	Not required	5–50	0.3–3.5	Economical energy. Limited to primary applications. High voltage source.	Sizes from 1.5 to 1,000 v.
AgCl	1.60–1.70	1.20–1.50			10–70	0.7–5.0		

NOTES:

* Operational temperature limits disclose a range at which the electrochemical system will function to deliver useful energy; all secondary systems are degenerated significantly at continuous operation above 100°F.

† Energy density is a function of discharge rate, temperature, design, and size of unit; except for remote-activated batteries, energy density values are based on typical cell capacities and overall volumes for sizes 0.5 ah upward.

Sizes and Applications

Silver oxide batteries are suited for hearing aids, instruments, photoelectric exposure devices, electric watches, and reference voltage sources. A cutaway of a silver oxide cell is shown in Fig. 48.

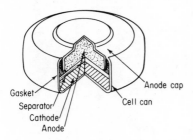

Fig. 48 Cutaway view of silver oxide cell. (*Eveready, Union Carbide Corporation.*)

Commercial forms of the zinc|potassium hydroxide|silver system, often referred to as the silver-zinc form, are produced as both primary and rechargeable cycle types but have only short cycle life.

Table 30 lists a group of typical silver oxide button cells, with their applications in electronic systems and instruments, ANSI and other designations, the measure of popular acceptance by those units with NEDA designations, the constructions, dimensions, weights, volumes, and, if given, competitive mercury equivalents.

Typical is the Gould Marathon rectangular form, shipped in a dry-charged condition. Cell cases are normally high-impact styrene. The negative plate (cathode) is sintered fine silver powder; the positive plate (anode) is activated zinc. The plates are separated by semipermeable ion-exchange membranes, while the electrolyte is potassium hydroxide. Battery containers for assemblage of cells are stainless steel, fiber glass, or plastics. Heaters may be installed within the battery container when close voltage regulation is required at low operating temperatures and may be thermostatically controlled to maintain desired temperature limits.

Wet life is the length of time the cell will operate after activation. It ranges from 48 hr to 18 months. Storage conditions for a dry-charged battery are: optimum 32 to 90°F, permissible −65 to 160°F (3 yr storage). When activated, operating temperatures are −50 to 160°F, but operation is usually 0 to 140°F. Sizes are given in Table 31, capacities in Fig. 49, and voltage in Fig. 50. Wilson[7] discussed sealed cadmium–silver oxide batteries.

There has been a need for a battery with high power-to-weight ratio and long cycle life. The zinc–silver oxide couple is one of the best as far

TABLE 30 Silver Oxide Button Cells

Voltage	Application	ANSI designation	Other designation	NEDA	Current, ma	Service, mah	Number of cells and sizes	Diameter in.	Diameter mm	Height in.	Height mm	Weight, oz	Volume cu in.	Volume cu cm	Mercury equivalent (approximate)
1.5	Hearing aids	.....	313	.....	5	30	1/313	0.310	7.87	0.140	3.55	0.02	0.01	0.15	
1.5	Hearing aids	S5	320	.....	5	60	1/SF	0.310	7.87	0.210	5.33	0.04	0.016	0.24	M5
1.5	Electric watches	S10	415	.....	0.1	100	1/S10	0.455	11.56	0.165	4.19	0.06	0.02	0.31	M10
1.5	Micro lamps	S10	415	.....	10	105	1/S10	0.455	11.56	0.165	4.19	0.06	0.02	0.31	M10
1.5	Hearing aids	S10	415	.....	10	105	1/S10	0.455	11.56	0.165	4.19	0.06	0.02	0.31	M10
1.5	Micro lamps	S15	421	.....	10	165	1/S15	0.455	11.56	0.211	5.34	0.08	0.03	0.47	M15
1.5	Hearing aids	S15	421	1107	10	165	1/S15	0.455	11.56	0.211	5.34	0.08	0.03	0.47	M15
1.5	Lightmeter	S16	421	1107	0.24	165	1/S16	0.455	11.56	0.211	5.59	0.09	0.03	0.47	
6	Electronic shutter and lightmeter	4S16	.....	1406A	10	165	4/S16	0.510	12.95	0.99	25.14	0.05	0.18	2.81	

TABLE 31 Battery Sizes

Cell-case dimensions (including terminals) A B C[h]	Low rate: SZR(L) Types Minimum cycle life: 25–50 cycles Wet life 12–18 months					High rate: SZR Types Minimum cycle life: 10–20 cycles Wet life 6–8 months				
	Cell types[a]	Max. discharge rate, amp[b]	Weight, lb[c]	wh/lb (initial)[f]	wh/cu in. (initial)[f]	Cell type[a]	Max. discharge rate, amp[b]	Weight, lb[c]	wh/lb (initial)[a]	wh/cu in. (initial)[a]
1.08 × 0.54 × 1.56	SZR-1LB	1	0.047	31.9	1.66	SZR-1HB	3	0.047	38.1	1.95
1.08 × 0.54 × 2.02	SZR-2LC	2	0.069	43.4	2.50	SZR-2HC	5	0.069	42.8	2.52
1.72 × 0.59 × 2.89	SZR-4LE	4	0.187	38.6	2.51	SZR-5HE	15	0.187	41.7	2.56
1.72 × 0.59 × 3.36	SZR-5LF	5	0.231	38.9	2.79	SZR-6HF	18	0.251	40.8	2.54
2.08 × 0.80 × 2.91	SZR-7LG	7	0.281	37.3	2.21	SZR-8HG	25	0.30	45.1	2.94
2.32 × 0.75 × 4.79	SZR-13LK	13	0.61	35.0	2.56	SZR-18HK	50	0.60	46.8	3.43
2.11 × 0.88 × 6.80	SZR-25LN	25	0.85	44.1	3.00	SZR-30HN	100	0.85	51.1	3.44
2.15 × 3.36 × 6.80[e]	SZR-25-4LN	25	3.25	46.1	3.05	SZR-30-4HN	100	3.35	51.9	3.52
2.06 × 1.74 × 4.53	SZR-25LP	25	0.90	41.6	2.41	SZR-30HP	100	0.90	51.6	3.00
3.23 × 0.89 × 7.02	SZR-30LS	30	1.06	42.0	2.30	SZR-40HS	125	1.20	59.5	3.66
3.23 × 1.01 × 6.85	SZR-40LU	40	1.43	41.9	3.73	SZR-55HU	125	1.50	61.0	4.10
5.30 × 3.23 × 6.40[a]	SZR-50-5LW	50	8.20	45.7	3.40	SZR-60-5HW	250	8.3	62.5	4.73
3.29 × 2.20 × 6.91	SZR-140LY	140	3.50	60.0	4.20	SZR-150HY	250	3.44	66.6	4.61

Cell-case dimensions (including terminals)[a]			Fast activating: SZFA Types Minimum cycle life: 2-5 cycles Wet life 30-75 days					Primary: SZMP Types Minimum cycle life: 1 cycle Wet life 48 hours				
A	B	C[h]	Cell type[a]	Max. discharge rate, amp[b]	Weight, lb[c]	wh/lb (initial)[a]	wh/ cu in. (initial)[d]	Cell type[a]	Max. discharge rate, amp[b]	Weight, lb[c]	wh/lb (initial)[d]	wh/ cu in. (initial)[a]
1.08 × 0.54 × 1.56			SZFA-1HB	3	0.061	36.3	2.20	SZMP-1.5HB	10	0.054	40.1	2.37
1.08 × 0.54 × 2.02			SZFA-2HC	5	0.073	47.8	2.74	SZMP-3HC	18	0.095	45.8	2.98
1.72 × 0.59 × 2.89			SZFA-6HE	15	0.208	46.2	3.14	SZMP-7HE	40	0.219	47.8	3.41
1.72 × 0.59 × 3.36			SZFA-7HF	18	0.231	47.1	3.19	SZMP-8HF	45	0.271	48.1	3.62
2.08 × 0.80 × 2.91			SZFA-10HG	30	0.406	49.0	4.20	SZMP-12HG	60	0.438	49.7	3.58
2.32 × 0.75 × 4.79			SZFA-25HK	50	0.656	66.0	5.28	SZMP-30HK	125	0.75	58.0	5.27
2.11 × 0.88 × 6.80			SZFA-35HN	125	0.85	59.7	4.06	SZMP-40HN	125	1.05	55.2	4.59
2.15 × 3.36 × 6.80[e]			SZFA-35-4HN	125	3.5	58.0	4.31	SZMP-40-4HN	125	3.8	61.1	4.71
2.06 × 1.74 × 4.53			SZFA-40HP	125	0.90	64.4	3.74	SZMP-45HP	135	1.13	61.9	5.17
3.23 × 0.89 × 7.02			SZFA-50HS	150	1.20	60.3	3.71	SZMP-55HS	150	1.45	71.4	5.31
3.23 × 1.01 × 6.85			SZFA-65HU	175	1.5	62.8	4.21	SZMP-75HU	200	1.77	74.4	5.90
5.30 × 3.23 × 6.40[g]			SZFA-70-5HW	300	8.3	61.2	4.61	SZMP-90-5HW[e]	350	9.35	76.3	6.51
3.29 × 2.20 × 6.91			SZFA-180MY	450	3.7	72.9	5.22	SZMP-220HY	500	4.03	79.9	6.47

[a] The numbers in the cell type designation indicate the nominal capacity of the cell, in ampere-hours. [b] To a cutoff voltage of 1.0 v. With cell filled ±3%. [d] Based on a plateau voltage of 1.45 v. [e] Available in 5 cell monobloc; figures given are for the entire monobloc. [f] Based on plateau voltage 1.50 v (10-hour rate). [g] Available only in 4 cell monoblocs; figures given are for the entire monobloc. [h] Overall terminal height can vary with application requirements.

93

as power-to-weight ratio is concerned, but it has a short cycle life. The nickel-cadmium couple has a long cycle life, but a low power-to-weight ratio. Since zinc is the component detrimental to the cycle life of the zinc–silver oxide couple, and silver oxide has a greater power output per unit weight than nickel, it can be seen that the cadmium–silver oxide couple should have both long cycle life and a high power-to-weight ratio. The cadmium|potassium hydroxide|silver oxide system is superior to the zinc–silver oxide system in cycle life and superior to the nickel-cadmium system in power-to-weight ratio.

In Fig. 51, a comparison of watthours per pound of two sealed D cells and the cadmium–silver oxide cell is shown. The cadmium–silver oxide cell is superior to the zinc–mercuric oxide cell at a 1-hr discharge rate. The cells are approximately equal at a 5-hr discharge rate. The zinc–mercuric oxide cell is superior at the 20-hr rate, but this cell is good for only one cycle.

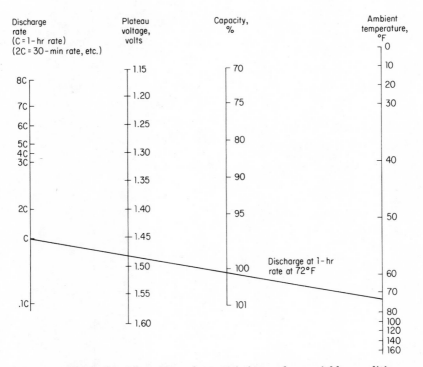

Fig. 49 Nominal performance characteristics under variable conditions. To find the capacity and the plateau voltage of a typical silver-zinc SZR-type cell, draw a straight line between the discharge rate and the ambient temperature. (*Gould Marathon Battery Company.*)

The Silver Batteries   95

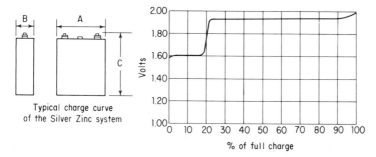

Fig. 50 The recommended charging cutoff voltage for silver-zinc models is 2.00 v/l. (*Gould Marathon Battery Company.*)

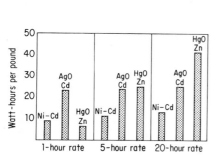

Fig. 51 Watthours per pound discharge characteristics of D-size cell. (*Eagle-Picher Industries, Inc.*)

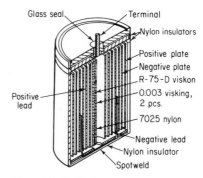

Fig. 52 Rolled-type construction. (*Eagle-Picher Industries, Inc.*)

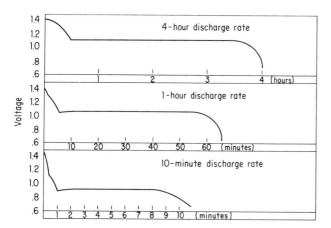

Fig. 53 Discharge characteristics at various rates. (*Eagle-Picher Industries, Inc.*)

The ideal separator material for this system is wettable, inert in the electrolyte, and impervious to oxidation and penetration by the silver oxide electrode material at both high and low temperatures. The ideal separator has not been found. This necessitates several types of separation.

The cells have been cylindrical D size, using the rolled-type construction as shown in Fig. 52.

Although the cadmium–silver oxide system has lower plateau voltages than the zinc–silver oxide or the nickel-cadmium systems, it has similar flat discharge characteristics. The voltage-versus-time curves for three different discharge rates can be seen in Fig. 53.

The system would be capable of replacing zinc–silver oxide cells for relatively high rate applications where a longer cycle life is needed, and it should be able to replace nickel-cadmium cells in applications requiring higher power output at some sacrifice in cycle life. The system is comparable to the zinc–mercuric oxide system and has the advantage of rechargeability.

REFERENCES

1. I. A. Denison, *Trans. Electrochem. Soc.*, **90**: 387–403 (1946).
2. W. Morrison, U.S. Patent 976,092 (1910).
3. H. G. André, U.S. Patent 2,317,711 (1943).
4. E. W. Jungner, U.S. Patent 692,298 (1902).
5. F. Jirsa, *Z. Elektrochem.*, **33**: 129–134 (1927).
6. E. M. Morse, Battery Experience on the Ballistic Missile Programs, *Bull. No. 106, National Winter Convention on Military Electronics*, sponsored by IRE (1962).
7. J. K. Wilson, Sealed Cadmium–Silver Oxide Batteries, *Bull. No. 105, Proc. 15th Annual Power Sources Conference*, 1966.

chapter 10
Water-activated Systems

Water-activated batteries satisfy an area of need for radiosondes, particularly meteorological, and for a variety of sea-associated emergency uses including lighting, signal beacons, and markers. Inherently the devices (1) have an active lifetime limited to hours, (2) should be maintained under load when activated, and (3) operate from a few tenths of watts, for lighting, to the output level of radio B batteries. Electrically, the batteries are noisy owing to chemical side reactions, but these same reactions provide the heat to make possible low-temperature and, therefore, upper-atmosphere performance. Energy density is relatively high because only active materials in a minimal package (without stored electrolyte) are involved. Values approximate 20 w/lb and 1.0 w/cu in.

Two chemical systems are used. Each involves a sheet magnesium anode and a water electrolyte with sea or salt water being preferred because of its higher conductivity, which sharply reduces activation time to a few seconds, whereas fresh water may require a few minutes. The difference is the choice of a silver chloride or cuprous chloride cathode. The basis for the choice of cuprous chloride is economic, and this reflects a balance of material and manufacturing costs.

Magnesium–Silver Chloride

A compact, simple, and efficient design of spirally wound cells was developed in batteries with capacity over about 5 amp-min. The cathode of each cell consists of silver foil, 0.001 to 0.003 in. (25 to 76 μ) thick, coated with electrolytically formed silver chloride on both sides, with wire leads welded on. The electrochemical equivalents of the deposits range from 0.75 to 5.0 amp-min/sq in. (0.12 to 0.78 amp-min/sq cm) depending on current, life, and output required in the end use. The anode is of magnesium foil of the same width to which a similar wire lead is welded. The two electrodes are wound spirally together, with dry absorbent paper interleaved between them, forming a cylindrical cell with wire leads at the ends. The cell thus formed in its dry condition is inert. The absorbent paper is a barrier between the electrodes. When saturated with an electrolyte, the absorbent paper becomes conductive, and the cell can be discharged.

For high-voltage applications, in which current and total output are low, the batteries are in flat-cell form. Area of these cells ranges from 0.59 to 2.625 sq in. (3.78 to 16.8 sq cm). In such cells the silver or copper chloride is placed only on one side of the sheet. Batteries are made up by stacking plates with absorbent paper between the electrodes of each cell and insulation between adjacent cells.

The batteries are manufactured at a temperature of 27°C (80°F) at a relative humidity of 20%, or an absolute humidity of 0.0045 lb water per pound of dry air (dew point of 3.33°C or 38°F). The batteries are packaged in sealed cans in this atmosphere with silica gel as desiccant. Under this condition, storage life is excellent.

At the time of use, the battery is removed from the sealed container and is activated by immersion in common or salt water to saturate the absorbent paper. It may be discharged either after removal from the water or while immersed continuously, i.e., in the case of low-voltage batteries.

With tap water as electrolyte, the low initial concentration of ions causes a comparatively slow rise in voltage for the first few seconds, but chloride and magnesium ions are rapidly produced by the cell action, which increases the conductivity of the electrolyte. Voltage ordinarily rises to nearly the peak within 1 to 2 min.

Specialized batteries have been developed for torpedoes and underwater missiles based either on the magnesium–silver chloride system or the zinc–silver oxide water-activated unit. Typical performance of the latter, as well as for reentry vehicles and spacecraft power supply, are given in Table 32.[1]

Aqualites have been used in the quick location of survivors of wrecks at

TABLE 32 Special Water-activated Units

Characteristics	Torpedo propulsion Model P293A	Reentry vehicle Model P212A	Spacecraft power supply Model P51A
Nominal voltage, v			28
Section I	76	28	
Section II	26	30	
Nominal capacity, ah			40
Section I		0.2	
Section II		0.6	
Specification capacity, ah			
Section I	78.75		
Section II	21.0		
wh/lb, actual	28.5		34
Maximum discharge rate, continuous, amp			300
Section I	525	13	
Section II	45	60	
Activation method	Automatic	Automatic	
Manually initiated			Semiautomatic*
Activation time, sec		0.5	
Section I	15		
Section II	6		
Activated stand capability, min		5	
Wet stand capability, days			60
Shelf life, years	2		
Unactivated		5	
Dry			3
Weight, lb	265	7.0	35
Wet, lb			44
Volume, cu in	5,360	126	700
Canister	Stainless steel	Hermetically sealed stainless steel, electro polished. Contains activation indicator	Welded stainless steel. Contains thermostatically controlled heaters and 4 psig pressure-relief system
Environmental capability:			
Operating			
Shock, gravities	60	200	100
Vibration, gravities	3	30	8
Temperature, °F	+28°–95°†	+25°–+160°‡	+0–+160°§
Acceleration, gravities		100	10
Altitude, ft		200,000	200,000
Nonoperating			
Shock, gravities		200	100
Vibration, gravities			5
Random, RMS		Equal to 21	
Temperature, °F		+65°–+160°	−35°–+125°
Acceleration, gravities		15	10
Altitude, ft		200,000	60,000

* Single-point activation accomplished with compressed air and A51A activation mechanism.
† Available without heaters to operate at ambient temperatures down to +60°F.
‡ The addition of heaters would allow operation at temperatures as low as −65°F.
§ Available without heater to operate at ambient temperatures above +40°F.
SOURCE: Whittaker Corporation, Denver, Colo.

sea. To this end a range of Aqualites and flashing beacons has been produced by the McMurdo Instrument Company Limited, Rodney Road, Portsmouth, Hampshire, England, to cover a wide variety of mishaps that occur at sea during the hours of darkness, or when darkness precedes the arrival of a rescue party.

Aquacels are chloride depolarized water-activated batteries made with one of two basic electrochemical systems, either silver chloride–magnesium or cuprous chloride–magnesium.

Magnesium–Cuprous Chloride

In general, the silver chloride system is preferable to cuprous chloride as it has better discharge characteristics over a wide temperature range. The cuprous chloride system may prove to be cheaper when large quantities of power are required. When Aquacels are to operate without gas evolution, e.g., to be completely sealed during discharge, the magnesium may be replaced by other metals.

Both of these batteries have indefinite storage life provided they are kept in a sealed condition. Once unsealed and immersed in water, they must be considered expended when all their power has been used up. It is not possible to recharge either of these electrochemical systems.

Sizes and Applications

High watthour-per-unit-weight ratios can be realized, 40 wh/lb (100 wh/kg).

They have been employed for life-jacket lights, life-raft lights, lifebuoy lights for air, surface, and underwater craft, signal flare initiators, radio distress beacon power supplies, flare-path markers, portable hand lamps, flashing beacons, and divers' and frogmen's lights.

For military equipment they become flare initiators, depth charge

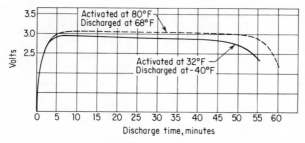

Fig. 54 Duration of discharge. (*The McMurdo Instrument Company Limited.*)

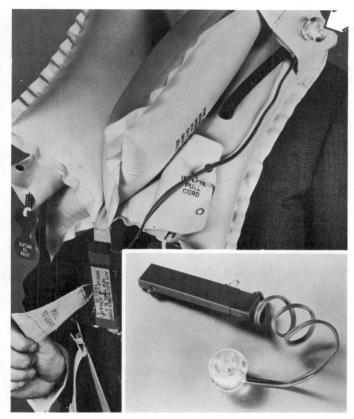

Fig. 55 A water-activated rescue light. (*ACR Electronics Corp.*)

initiators, signal lights, marker lights, underwater lights for swimmers and divers, glove warmers for underwater swimmers, and radio power supplies.

A typical application for meteorological balloons is built with a connected lamp. Duration of discharge is 45 min minimum with 0.75 w nominal lamp (2.5 v, 0.3 amp, see Fig. 54). Lamp output is 12.5 lumens average at 2.9 v. Overall size is $1\frac{3}{4} \times 2\frac{1}{2} \times \frac{3}{4}$ in. ($45 \times 64 \times 20$ mm) maximum. Weight in operation is 0.9 oz (25 g) maximum.

A water-activated rescue light is shown in Fig. 55. The battery is $4\frac{3}{4} \times 1\frac{3}{8} \times \frac{3}{8}$ in., weighing $1\frac{3}{4}$ oz complete with encapsulated bulb of 1.5 w and 14 in. of cable, giving a light intensity of 1 cp for 8 hr. The battery is a magnesium–silver chloride unit with a dry-storage life of 5 years.

chapter 11
Obsolete and Historical Systems

Before the expansion of electrical transmission lines and extensive distribution systems, a large number of different primary batteries were marketed. Some of these were employed only in laboratory work, military activities, army field-communication systems, or isolated telephone exchanges. Over the years the following systems have achieved prominence:

1. The Bunsen, with the system $HgZn|1$ part $H_2SO_4 + 12$ parts $H_2O|$concentrated $HNO_3|C$, voltage—1.94 v
2. The bichromate or Poggendorf, with the system $HgZn|1$ part $H_2SO_4 + 12$ parts $H_2O|$concentrated solution $Na_2Cr_2O_7 + H_2SO_4|C$, voltage—2.00 v
3. The Daniell or gravity, with the system $HgZn|5\%$ solution $ZnSO_4 \cdot 6H_2O|$saturated solution $CuSO_4 \cdot 5H_2O|Cu$, voltage—1.07 v
4. The Féry, with the system $Zn|12\%$ NH_4Cl solution|depolarizing C, voltage—1.2 v
5. The Grove, with the system $HgZn|1$ part $H_2SO_4 + 12$ parts $H_2O|$fuming $HNO_3|Pt$, voltage—1.96 v

6. The Leclanché, with the system HgZn|20% NH$_4$Cl solution MnO$_2$|C, voltage—1.5 v

While interesting historically, these systems are commercially obsolete. The last in its modern dry form is the "dry cell."

The Daniell cell had the system Zn|ZnSO$_4$|CuSO$_4$|Cu,

$$Zn + CuSO_4 \rightleftharpoons ZnSO_4 + Cu$$

whose voltage is given approximately by the formula

$$E = 1.1 + 0.029 \log \frac{Cu^{++}}{Zn^{++}} \quad v$$

where the ionic ratio is that of the molar Cu^{++} concentration of the catholyte to the molar Zn^{++} concentration in the anolyte. If there be little difference in the concentrations of the copper and zinc sulfates, the ionic ratio can be replaced by the ratio of the two salts whose ionizations are quite similar under similar conditions. The emf of the cell is independent of temperature. The sign of the temperature coefficient is a function of the concentration of the electrolytes, becoming more positive with increase of the ratio Cu^{++}/Zn^{++}.

The commercial form was a glass jar which held an inner porous cup. The cup was surrounded by a saturated solution of CuSO$_4$ containing excess crystals and further by the cathode of copper foil bent to a cylindrical form. The inner pot contained ZnSO$_4$ solution acidified with H$_2$SO$_4$, into which dipped the amalgamated zinc anode rod. The cell emf was a function of the concentration of the ZnSO$_4$ solution, having a maximum value of about 1.14 v. Without the addition of H$_2$SO$_4$ the voltage became 1.07 v. The setup is not adapted to stand on open circuit, since the chemical reactions between the constituents continue whether the cell be in use or not.

The Lalande "caustic soda primary battery" had an electrolyte of caustic soda solution. Amalgamated zinc formed the anode and cupric oxide the cathode and depolarizer. A thin layer of oil on the surface of the electrolyte prevented evaporation and the formation of carbonate. The copper oxide and zinc electrodes with insulators were arranged so that the entire assembly could be secured to a cover, usually of porcelain, by a suspension bolt and nuts and washers. The terminal wire from the zinc was a copper wire with insulation of a rubber compound.

More than 90% of all the batteries of this type were used for railroad signals and other railroad devices. The units contained 4,400 cc of water in which the NaOH was dissolved. The system is Zn|alkali solution|oxides of copper|Cu. A typical cell is shown in Fig. 56. The containing jar was made to withstand the corrosion of the electrolyte and

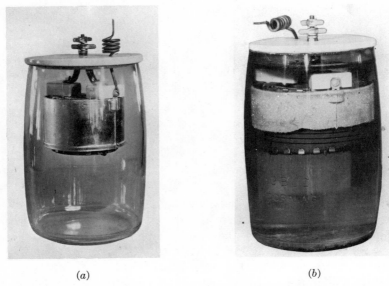

Fig. 56 (a) Lalande cell. (b) Exhausted Lalande cell. (*The Waterbury Battery Company.*)

resisted cracking due to the heat generated in dissolving the caustic soda. The electrolyte was 18 to 19% NaOH. The CuO, which functioned as cathode and depolarizer, was made on a compressed or loose form. The CuO was ground to a fine powder, mixed with a suitable binder, and subjected to pressure. It was baked and the outer surface metallized by partial reduction to increase the conductivity. Copper oxide is a poor conductor. The container for the loose oxide, as well as the finished form of the compressed oxide, took the shape of a flat plate or hollow cylinder. The zinc electrode was cast in a cylindrical form and amalgamated with mercury either in the casting or by dipping in a mercury bath, the mercury content being about 2.5%. The Edison modification of the Lalande cell uses flat-plate electrodes, the zinc plates being cast with ribs. The compressed copper oxide electrode was suspended centrally with zinc on each side.

The chemical reaction of the unit may be represented by the equation

$$Zn + 2NaOH + CuO \rightleftharpoons Na_2ZnO_2 + H_2O + Cu$$

The emf is approximately 0.95 v, but with heavy currents, the terminal voltage dropped to 0.6 v. The internal resistance of the cell was low. During operation the CuO electrode was reduced to copper. Upon exhaustion, this electrode had to be washed and reoxidized by heating at 150°C, and fresh zinc plates were added. Commercial batteries ranged from 75 ah up to cells with a rated capacity of 1,000 ah. Seven unit electrode systems in a single container were made for batteries which maintain their voltages from 500 to 1,000 ah.[1]

The Leclanché cell was $Zn|NH_4Cl|MnO_2|C$. In the original form the carbon rod was contained in a porous cup filled with crushed carbon and MnO_2. The mixture of these two was pressed to obtain intimate contact. In later forms the MnO_2 and carbon were molded together into cylinders by a binder, the zinc electrode suspended centrally. The electrolyte was a 20% NH_4Cl solution. The open-circuit emf of the cell was about 1.5 v, but when heavy currents were drawn from the cell, the terminal voltage dropped rapidly to 1.1 to 1.2 v. It is suitable for work on open circuits, but on closed circuits it is suitable only if large currents be drawn intermittently for short periods.

The AgCl cell, which is similar to the Leclanché save that AgCl is used as a depolarizer with a silver electrode, was made in small sizes for army maneuvers and field service. The life of this cell was rather long for the size, and the open-circuit voltage is 1 v. With magnesium substituted for zinc, the system became the basis for cells widely used in World War II.

The perchloric acid cell contained plated lead dioxide positive plates, metallic lead negative plates, and aqueous perchloric acid electrolyte. The soluble nature of the discharge product, lead perchlorate, permitted relatively high currents to be drawn from the cell at temperatures as low as $-20°C$. Discharge data at current densities from 0.045 to 3.33 asi (0.70 to 51.6 asd), temperatures from -40 to $+40°C$, and acid concentrations from 40 to 73% are given by White and his associates.[2]

The perchloric acid cell was developed by Schrodt, Craig, and Vinal[3] for small lightweight batteries for radiosonde equipment. The overall reaction during discharge has been represented as follows:

$$PbO_2 + Pb + 4HClO_4 \rightleftharpoons 2Pb(ClO_4)_2 + 2H_2O$$

Acid efficiencies at each concentration based on the equation approach 100% as the current density is reduced. Efficiencies greater than 100% at low rates and high temperatures indicate a two-stage discharge process under these conditions with the formation of basic perchlorates as reaction products.

Positive plates were prepared by plating a dense deposit of lead dioxide on an inert conducting grid. Lead dioxide can be anodically plated from a lead nitrate bath.[4]

The powerful oxidizing conditions accompanying the plating of lead dioxide anodically also tend to dissolve most anode bases. Schrodt, Craig, and Vinal[3] employed palladium as a positive grid material, since the plates were tiny and did not require large quantities of the precious metal. The cost of this material for larger and higher-rate cells would be prohibitive, however, since large grid areas would be required. The application of nickel as a base for lead dioxide anodes[5] suggested its use for positive grids. Nickel-plated aluminum, iron, or copper have been used.

Discharge characteristics of the perchloric acid cell depend on current density, acid concentration, and temperature. Commercial 71% perchloric acid is an electrolyte with good efficiency at temperatures of 25°C or higher and low current densities. As the current density is increased or the temperature lowered, the acid must be diluted.

The perchloric acid cell has limitations for general use. No positive grid material other than the precious metals has been found which will resist attack in contact with perchloric acid and lead dioxide for more than a few days. Cells using nickel grids must not be filled with acid until immediately before discharge. The dry form in storage should last indefinitely.

REFERENCES

1. M. L. Martus, *Trans. Electrochem. Soc.*, **68**: 151 (1935).
2. J. C. White, W. H. Power, R. L. McMurtrie, and R. T. Pierce, Jr., *Trans. Electrochem. Soc.*, **91**: 73–94 (1947).
3. J. P. Schrodt, D. N. Craig, and G. W. Vinal, Report of the National Bureau of Standards to the Navy Department, Bureau of Ships, *NBS Project 1911* (October, 1943).
4. F. Ferchland and J. Nussbaum, U.S. Patent 900,502 (1908); Y. Kato, British Patent 456,082 (1935).
5. Y. Kato and K. Koizumi, *J. Electrochem. Assoc. Japan*, **2**: 309 (1934); Y. Kato, K. Sugino, K. Koizumi, and S. Kitahara, *Electrotech. J.* (Japan), **5**: 45–48 (1941).

chapter 12
Reversible Systems—Secondary Cells

Cells which are reversible to a high degree, in that the chemical conditions may be restored by causing current to flow into the cell on charge, are classed broadly as storage batteries or electric accumulators. The form in widest commercial use is the Pb-H_2SO_4 type. Another prominent form in the United States is the Ni-Fe-caustic cell, or "Edison Cell."

In 1859 Planté devised a cell for the storage of electrical energy consisting of two sheets of lead separated by strips of rubber, rolled into a spiral form, and immersed in a dilute (about 10%) solution of H_2SO_4. He found it possible to increase the capacity of the cell materially by "formation." Following periods of charge, he discharged the cell or allowed it to rest, during which time local action transformed the covering of peroxide on the positive plate into $PbSO_4$. From time to time he reversed the polarity and repeated the process of charge and discharge to build up the capacity of the cell.

The positive pole of a cell is that one from which the current flows into the external circuit. In storage-cell practice, a positive plate is one which is connected to the positive pole, and the negative plate is one which is connected to the negative pole.

TABLE 33 Comparison of Common Battery Systems Rated at 100 Ampere-hours, 12 Volts

Type	Average volts per cell	Number of cells (12 v)	wh/lb	wh/ cu in.	Charging time	Estimated life	Estimated cost, dollars		
							Total	Per year	Per cycle
Lead-acid industrial	1.9	6	13.1	1.35	5–10 hr at constant potential or stepped current	1,600 cycles 5–10 years	70	9	0.043
Automotive	1.9	6	15.0	1.3		300 cycles 3 years	40	13	0.13
Nickel-cadmium pocket	1.2	10	11.0	0.6	7–8 hr at constant potential	2,000 cycles 10–20 years	150	6	0.075
Sintered	1.2	10	11.5	0.88	2 hr if vented, 10 sealed	3,000 cycles 10–20 years	300	20	0:10
Nickel-iron (tube type)	1.2	10	10.6	0.92	6–7 hr at constant potential	2,000 cycles 15–20 years	130	7.5	0.065
Zinc-silver	1.55	8	85–100	3.0	4–20 hr at constant potential	200 cycles 1 year	800	800	4.00
Cadmium-silver	1.06	11	50–75	2.5	4–20 hr at constant potential	500 cycles 2–3 years	1,000	400	2.00
O_2–H_2 fuel cell	0.7	16	......	......	......	......	......	......	......

........not applicable or available.

For extended commercial use a storage cell should have high capacity per unit of weight. Chemical effects that cause deterioration or loss of stored energy should be absent. The transformation of electrical into chemical energy as in charging, and of chemical into electrical energy as in discharging, should proceed nearly reversibly. An ideal storage cell should have low resistance, have simplicity and strength of construction, be durable, and be producible at low cost. In the past a large number of storage-cell constructions and systems have been proposed. Almost all of them had disadvantages which prevented their commercial adoption. The $Pb-H_2SO_4$ type is in universal use. It does not suffer from deteriorating chemical effects, shows almost reversible transformation of electrical into chemical energy and vice versa, has low resistance, can be used for long periods with proper care, and has low first cost. It does not have as high capacity per unit of weight as might be desired; and while its construction is simple, the mechanical strength of the lead plates is low. On the other hand, the Ni–Fe–caustic potash cell shows high capacity per unit of weight, absence of deteriorating chemical side reactions, and ability to withstand long, continued use. Its first cost, owing to expensive materials and complex construction, is high, but the cell is mechanically strong and durable. A disadvantage is that it shows charging and discharging losses owing to the incomplete reversibility of the electrical and chemical energy transformations in the system.

Other systems seeking their place in the industrial sunshine of rechargeable cells are the nickel-cadmium, zinc-silver, and zinc-cadmium systems, particularly in portable electric energy sources. In the myriad of new uses criteria such as energy/weight ratio may be all important; in other applications criteria may be cycle life, shelf life, energy/volume ratio, or cost.

A comparison of common secondary batteries, rated at 100 ah, 12 v, is given in Table 33. It is not readily possible to calculate the cost of power, as in essence it is recovered from the cost of power charged into the cell at the central station householder's rate of 1.2¢ per kwh, corrected for the efficiency of the cell. It is apparent that more energy is charged into the cell than is recovered therefrom. In addition the battery has a depreciation in that its useful life is not infinite. As a structure it may have a scrap value at the end of its useful life, at least for its lead metal content. A 100-ah, 12-v cell might be considered a 1-kwh unit and its power cost per kwh is on the order of 100 to 1,000 times the householder's central station rate.

chapter 13

Lead Secondary Cells

Theory

The Pb-H_2SO_4 cell system is $PbO_2|H_2SO_4|$sponge Pb. On discharge, both the peroxide on the positive plate and the lead on the negative plate are quantitatively converted into $PbSO_4$ according to the reactions

$$PbO_2 + H_2SO_4 \rightleftharpoons PbSO_4 + H_2O + O$$
and
$$Pb + H_2SO_4 \rightleftharpoons PbSO_4 + 2H$$

which may be combined into the reaction

$$PbO_2 + Pb + 2H_2SO_4 \rightleftharpoons 2PbSO_4 + 2H_2O$$

which, when read from left to right, is the equation of discharge and, inversely, the reactions during charge.

The reaction is that of the double-sulfate theory capable of thermodynamic proof. Although the theory gives us an equation for calculating the performance of the cell, it does not tell us about the actual processes taking place at the positive and negative plates. Discharge and charge reactions taking place in the cell are shown in Fig. 57.

A lead-acid storage battery is a simple thing. A "laboratory model" can be readily made by taking two strips of lead and hanging them in and on a glass jar and filling the jar with dilute sulfuric acid. A source of direct current is connected to these strips or "plates," allowing them to "charge." In a short time the surface of one will become dark brown in color while the other will retain its original lead color. The brown plate has become covered with a layer of lead peroxide and is the positive plate of the cell. The unchanged plate is the negative. When the dc charging source is removed, a sensitive voltmeter will indicate a voltage of approximately 2 v between the "terminals" of the cell. If an electrical load be connected to the terminals, a current will flow from positive to negative and the cell will deliver power. The thickness of this surface film, and therefore the cell's "capacity," can be somewhat increased by alternate "cycles" of charge and discharge.

Such a cell has no practical value because the available surface area of these two lead strips is not large enough to accumulate sufficient of the active materials, the brown lead peroxide of the positive, and metallic "sponge" lead of the negative. The problem in the development has been to increase the effective area of plate surface to achieve greater capacity for industrial use.

If the surface of the smooth lead plates be scratched or serrated or otherwise increased in effective area, the capacity of the cell is improved. More elaborate means of "roughening" the surface have resulted in commercial types. Plates have been cast with complex ridges or grooves and mechanically furrowed to obtain greater surface; also separate corrugated lead ribbons have been rolled into spiral "buttons" and inserted in lead alloy frames (see Fig. 58). Plates of these constructions have been used primarily as "positives" and give long life. Such plates, in which the active material is developed electrochemically from metallic lead, are Planté types.

Manufacture and Testing

Lead storage batteries are made in a wide range of sizes and capacities, from small elements up to units of very high capacity, weighing several tons, for central station work. For stationary cells, the containing vessels for the smaller and medium units are of glass; for the larger ones, lead-lined wood or plastics. Containers for batteries up to several thousand ampere-hours capacity are of glass.

The active materials, the PbO_2 on the positive plate and the sponge lead on the negative, are crystalline in structure, and the intergrowth of the crystals holds the masses together. It is probable that the positive active material is a hydrated peroxide of lead as it exists in the cell.

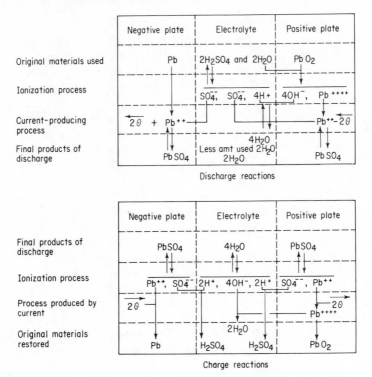

Fig. 57 Discharge and charge reactions of lead cell.

Many modifications have been proposed and are in use either for forming the active materials in place or for applying them and holding them in place mechanically. The plates may be of the Planté type comprising a mass of lead, of flat form, with a highly developed surface on which the active material is electrochemically formed as a coherent layer; or they may be pasted plates in which the active materials are cemented masses supported in a grid, usually of lattice form.

Planté plates are prepared from lead blanks which have been cast, rolled, cut, and stamped. A number of different methods have been used for increasing the surface of these plates. In one the plate is "plowed" so that fine furrows are made and leaves of lead thrown up on the plate. In others, the blanks are swaged by blocks having designs cut in their surfaces so that recesses are produced in the finished plate. In another case, the plates are spun to increase the surface. Highly developed surfaces have been obtained by casting. In the Manchester plate (Fig. 58), heavy grids of antimonial lead were cast with a large number of round holes into which buttons of soft lead with corrugated surfaces were pressed, the buttons being prepared from lead ribbon. The holes are made with a slight bevel so that, as the lead button grows during the operation of the cell, it becomes more and more tightly locked in the supporting grid. This construction is now too costly for commercial cells.

The active materials of the plate are obtained by oxidizing the surface of the lead plate or reducing this material to sponge lead.

The common method, however, of obtaining large "areas" of active material is to use very finely powdered lead oxides made up into "pastes." These are a sponge with the electrolyte filling all the pores and in contact with the active material over an area many times the evident surface of the pastes. In addition to the essential oxides these pastes contain other materials to make them cohesive, so that they do not wash away in the electrolyte, but at the same time porous, so that the electrolyte can circulate through them.

The active materials alone have no rigid form or strength and, particularly the positive, are poor conductors. It is necessary to mount them in lead alloy frames or "grids" to achieve and retain shape and to conduct the current to all parts of the material. This lead grid is usually a lattice work into which the paste is pressed, or a series of spines or "core rods," each surrounded by a perforated rubber, plastic, or glass fabric tube with the active material in the annular space between. The lattice type (Fig. 57) is commonly known as a "flat-plate" type. This construction is nearly always used for the negative plates and can be used for

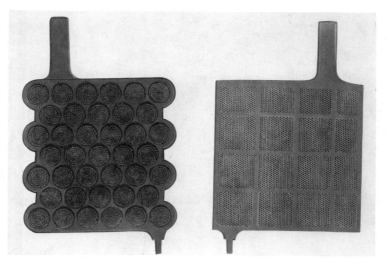

Fig. 58 A Manchester-positive and a box-negative plate. (*ESB Incorporated.*)

positives also. The spine-and-tube construction (Fig. 59) or a "tubular plate" is used only for positives.

In a fully charged battery all the active material of the positive plates is lead peroxide, and that of the negative plates is pure sponge lead. All the acid is in the electrolyte and the specific gravity is at maximum. As the battery discharges, some of the acid separates from the electrolyte which is in the pores of the plates, changing the active material to lead sulfate and water. As the discharge continues, additional acid is drawn from the electrolyte and further sulfate and water are formed. The specific gravity of the electrolyte will gradually decrease because the proportion of acid is decreasing and the water is increasing.

When a battery is placed on charge, the reverse takes place. The acid in the sulfated active material of the plates is driven back into the electrolyte, reducing the sulfate in the plates and increasing the specific gravity of the electrolyte. The specific gravity will rise until all the acid is driven out of the plates and back into the electrolyte. There will then be no sulfate in the plates.

After all the acid is returned to the electrolyte, additional charging will not increase the gravity. The acid in the cells is in the electrolyte, and the battery is fully charged. The material of the positives is again lead peroxide, the negatives are sponge lead and the specific gravity is at a

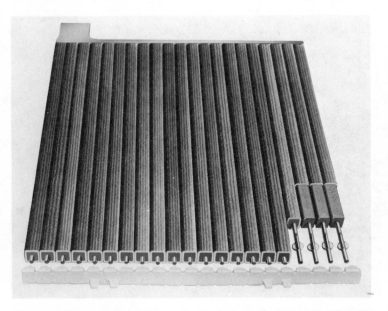

Fig. 59 The spine-and-tube construction or "tubular plate" grid. (*ESB Incorporated.*)

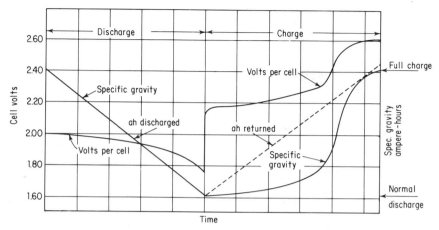

Fig. 60 Typical voltage and gravity characteristics during a constant rate discharge and recharge. (*ESB Incorporated.*)

maximum. On discharge, the plates absorb acid and on charge they return the acid to the electrolyte.

As the cells approach full charge they cannot absorb all the energy from the charging current. The excess electrolyzes water into hydrogen and oxygen liberated from the cells as gases. This is the reason for the addition of water to batteries.

The decrease in specific gravity on discharge is proportional to the number of ampere-hours discharged. This is shown by the straight line in Fig. 60. On recharge, however, the rise in specific gravity is not uniform or proportional to the amount of charge.

In the original process the plates were alternately charged and discharged with occasional reversals of the charging current until the plates had acquired sufficient capacity. The method requires considerable time and a large amount of electrical energy. The usual method involves "forming" agents which attack the lead of the plate. These are nitrates, chlorates, perchlorates, chlorides, fluorides, oxidizing agents such as bichromates and permanganates, and reducing agents such as formic acid, oxalic acid, alcohol, hydroxylamine, and sulfurous acid. These additional agents are confined to the positive plates which are the anodes in the forming bath. Negatives are obtained by a reversal of the positive plates during which the PbO_2 on their surfaces is reduced to sponge lead.

Pasted plates (Fig. 61) consist of grids or cast antimonial lead forms in the openings of which a paste, made by mixing litharge (PbO or red lead —Pb_3O_4—or a combination of these) with a dilute solution of H_2SO_4, is placed. It is probable that these two reactions occur:

$$Pb_3O_4 + 2H_2SO_4 \rightleftharpoons 2PbSO_4 + PbO_2 + 2H_2O$$
$$PbO + H_2SO_4 \rightleftharpoons PbSO_4 + H_2O$$

In addition, it is probable that hydration of the lead oxide also occurs.

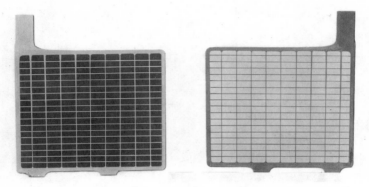

Fig. 61 Pasted positive and negative plates. (*ESB Incorporated.*)

Expanders of inert materials, lampblack, $BaSO_4$, graphite, and forms of carbon are added in amounts of a fraction of a percent to the materials for negative plates. They prevent the contraction and solidification of the sponge lead.

The pasted grids are allowed to dry and harden and are then electrolytically oxidized and reduced in either dilute H_2SO_4 or a sulfate solution, in lead tanks or in glass or rubber vessels holding a large number of plates. The positives and negatives are usually formed together for several days, although the latter require somewhat more time than the former. After this time the plates are washed and assembled in cells.

The elements are in the form of a thin paste with a grid structure of lead-antimony or silver-lead alloy, or in special forms of 0.4% calcium-lead. The alloy has physical strength and rigidity and offers resistance to "formation" or corrosion by the acid. The plates are arranged parallel to each other, alternately positives and negatives. All the positives are joined and connected by an alloy strap, as are the negatives. This strap, through its post, leads to the external circuit.

The length, width, thickness, and number of plates are determined by the required capacity. There is a negative plate at each end of the assembly. Thus a 15-plate cell has seven positive and eight negative plates. The two outside negative plates are frequently thinner. The plates are prevented from contact by a separator in microporous sheet forms of wood, rubber, glass, or plastic.

The plates and separators are in a container which holds the electrolyte. A cover is sealed to the top of the jar to exclude dirt or foreign material and reduce the water evaporation. The cover has a vent plug.

The lead-acid cell has the highest voltage of any commercial type. It has a nominal voltage of 2 v, although this varies with the specific gravity. Thus a three-cell battery is a 6-v battery, or a 60-cell battery is a 120-v battery, etc.

Lead-type secondary cells are marketed either charged and filled with electrolyte ready for service, or charged and dry, that is, without electrolyte, so that addition of H_2SO_4 solution of the required specific gravity is made when the battery is placed in service.

Exide has developed the Ironclad construction as shown in an assembled battery in Fig. 62, with some details in Fig. 63. The lead-antimony alloy has been replaced by a lead-silver alloy or lead-antimony silver Silvium. Active material is packed in fiber glass tubes or sleeves and assembled together into a plate as shown in Fig. 59. With more than a dozen in every plate, they hold the active material tighter and press it more firmly against the grid spine than any flat plate can. Assembly in the case is shown in Fig. 64. The separators are microporous rubber and have a ribbed structure and act as a retainer for the negative plate active material. The other plate construction is shown in Fig. 65.

In the charged and dry form, the chemical composition of the active materials on both the positive and negative plates is in that state of oxidation which corresponds to the charged condition. Plates must be

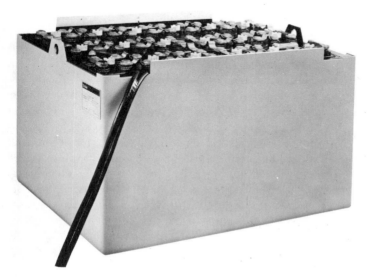

Fig. 62 Exide Ironclad battery. (*ESB Incorporated.*)

Fig. 63 Cutaway of ironclad battery construction. (*ESB Incorporated.*)

treated to render them stable during storage. Both the positive and negative plates start out according to normal procedures which involve electrolytically oxidizing lead sulfates to lead dioxide for the positive plate, and electrolytically reducing lead sulfates to porous metallic lead for the negative plate. In the case of the positive, the plates are drained of the H_2SO_4 used in the electrolytic oxidation, washed, and dried at 160 to 230°F. The positive plate is stable during storage.

For the negative plate, since residual H_2SO_4 acts as a catalyst for oxidation of the sponge lead, almost complete washing and removal of the forming (sulfuric) acid are required. Since moisture also catalyzes oxidation of the porous lead, the drying of the plate must be accomplished rapidly under conditions which exclude oxygen. One process dries the plates in an atmosphere containing little or no free oxygen by burning a gaseous fuel with air and using the heat generated to dry the plates while the products of the combustion surround the plates. The plates can be dried with a minimum of oxidation. After cooling, the plates are stable against oxidation under storage. An alternative procedure is to dry the plates rapidly under a high heat so that the steam generated presents a protective atmosphere.

Pasted plates are mechanically weaker than Planté plates, and the active material tends to fall out in time, particularly at high current density. One manufacturer employs positive plates with a paste of high capacity, with a spun-glass retainer of a porous nature and low resistance. The glass layer prevents the active material from falling out of the plates. Another maker has developed a pasted plate which is very hard, of high capacity, and porous. A slotted, thin, tough, hard-rubber retainer is on both sides of the positive plate. This retainer, in addition to the wedging of the elements in the rigid container, furnishes mechanical support. The capacity of a storage cell is stated in ampere-hours at some normal rate of discharge, the 8-hr rate being standard. The capacity of a cell with a definite type and thickness of plate is in proportion to the plate area. The emf or open-circuit voltage depends upon its chemical constituents. It varies further with the strength of the electrolyte, with the temperature, and to a minor extent with the state of charge of the plates, internal resistance of the cell, polarization, and acid concentration effects. Per ampere-hour of discharge, the amount of active material converted into $PbSO_4$ is 0.135 oz sponge lead and 0.156 oz PbO_2 independent of the rate of discharge. The amount of active material present in the plate is some three to six times that which under normal

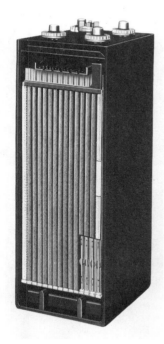

Fig. 64 Ironclad construction. (*ESB Incorporated.*)

Fig. 65 Iron plate construction. (*ESB Incorporated.*)

discharge of the cell is converted into $PbSO_4$. Part of this excess is present to give long life.

The open-circuit voltage of a lead cell varies from 2.06 to 2.14 v, according to the strength of the electrolyte and the temperature, and may be calculated from the formula

$$E = 1.850 + 0.917(G - g)$$

where G is the specific gravity of the electrolyte and g the specific gravity of water at the cell temperature.

In another form

$$\text{Volts} = \text{specific gravity} + 0.84$$

Therefore the open-circuit voltage of a cell with a specific gravity of 1.210 will be 2.05 v; one with a gravity of 1.280 will be 2.12 v.

On discharge, there is a decrease in voltage owing to the internal resistance. This "drop" increases with an increase in discharge current. At a continuous rate of discharge, the voltage becomes lower as the discharge progresses. As the cell nears exhaustion, the voltage drops very rapidly to and below an effective value of "final" voltage. It varies with the rate of discharge, being lower with higher ampere rates. It may be as high as 1.85 v for low rates or as low as 1.0 v at high discharge rates. A value of 1.75 v is representative of typical applications.

When a discharged battery is charged, its voltage rises, increasing with the charging rate. With average rates, the voltage will rise within minutes to 2.10 or 3.15 v, and increase gradually until the charge is perhaps three-quarters complete. Then the voltage rises more sharply and levels off at a maximum of 2.6 v.

The gravity must be high enough for the electrolyte to contain a sufficient amount of actual sulfuric acid to fulfill chemical requirements. If the gravity be too high the acid content may be strong enough to exert chemical attack. The full-charge gravities are:

1.275	Heavily worked or "cycled" batteries such as electric industrial trucks.
1.260	Automotive service.
1.245	Partially cycled batteries such as railway-car lighting and large engine starting batteries, etc.
1.210	Batteries in stationary standby or emergency service.

The specific gravity at any particular time is an indication of charge. A cell reads 1.195. This type has an average full-charge gravity of 1.245 and a gravity drop of 125 points (8-hr rate). It is 50 points below full charge and is $50/125$ or 40% discharged.

Typical charge and discharge curves are given in Fig. 66. It is often

Lead Secondary Cells   121

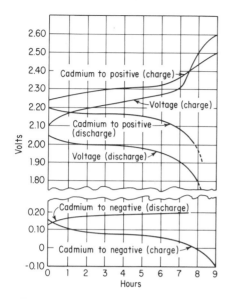

Fig. 66 Typical charge-discharge curves of lead storage battery.

Fig. 67 Cast-on-strap machine. (*Globe-Union Inc.*)

desirable to determine the relative performance of positive and negative plates in a cell. This may be done by taking the voltage between either group and a reference electrode such as zinc, sponge lead, or preferably cadmium. Cadmium curves are included in the diagram.

Battery assembly lines are highly automated. Figure 67 shows a Globe-Union cast-on-strap machine. Figure 68 is an element stacker and Fig. 69 a connection welder.

Vinal, Craig, and Snyder studied the addition of small amounts of cobalt, nickel, and iron sulfates to the electrolyte of commercial batteries which was then subjected to 150 cycles of charge and discharge. Cobalt decreased the charging potential of positive plates, and nickel decreased the charging potential of negative plates. Iron showed no significant effect on the potentials of either plate. Cobalt reduced the corrosion of some of the positive plates but had a harmful effect on the wood separators. Decreased positive plate potentials are of interest because they permit the use of lead alloys free from antimony.

As water is "consumed," the level falls and the remaining electrolyte has a higher specific gravity.

The higher the discharge rate in amperes, the less total ampere-hours

Fig. 68 "Reed" element stacker. (*Globe-Union Inc.*)

Fig. 69 HV connection welder. (*Globe-Union Inc.*)

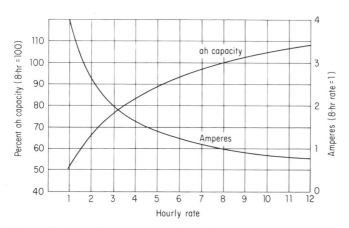

Fig. 70 Capacity-rate curve based on 8-hr rate. (*ESB Incorporated.*)

a battery will deliver under similar conditions. This relationship will vary somewhat with different types of plate and cell construction, but Fig. 70 shows such a relationship.

During discharge, the only portion of the electrolyte which is "useful" is that in the pores of the plates in contact with the active material. As the acid becomes depleted or exhausted, the electrolyte must diffuse or circulate to bring more acid to the active material. The higher the rate of discharge, the more rapid this circulation must be to maintain normal cell voltage. As the rate increases, this diffusion does not increase appreciably. The electrolyte in the pores of the plates is less dense. The cell voltage decreases and limits the total capacity.

The standard is the 8-hr rate which can be expressed, for example, either as 100 ah at the 8-hr rate or as 12.5 amp for 8 hr. Automotive batteries are rated at the 20-hr rate and motive power types on a 6-hr basis. Manufacturers usually list several hourly ratings including the 8-hr.

Chemical reactions are accelerated at higher temperatures. Resistance and viscosity of the electrolyte are reduced at higher temperatures, thus reducing the voltage drop and maintaining its terminal voltage at a higher value. These increase the capacity at higher temperatures and reduce it at lower temperatures. This difference in capacity varies with the rate of discharge as well. Figure 71 shows this approximate relationship.

The "standard" reference temperature is 77°F (25°C).

Electrolyte of higher gravity has more acid in contact with the active material and available for the chemical reactions than electrolyte of lower gravity. With a given total acid requirement, the need is met more readily by high gravity and with less rapid diffusion or circulation.

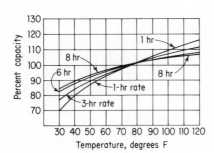

Fig. 71 Effect of temperature on capacity, 77°F = 100%. (*ESB Incorporated.*)

A difference of 25 points in gravity will change the capacity 8 to 10%. If a cell has 100 ah with a gravity of 1.275, its capacity will be 90 to 92 ah if the gravity is 1.250.

The "final voltage" depends on application. At low rates of discharge

(72-hr rate), this may be as high as 1.85 v per cell, whereas for the very heavy currents required for engine starting, it may be as low as 1.0 v. There is a nominal standard of 1.75 v per cell.

Battery rating must be qualified in many respects to performance. It might be specified as:

One 12-cell battery with a capacity of 500 ah at the 8-hr rate, at a temperature of 77°F, to a final voltage of 1.75 v per cell, full-charge specific gravity 1.250.

Goldberg and Reed of the Naval Ordnance Laboratory proposed a 0.9-v "dry cell" lead battery of 1.5 ah to power the electrochemical control units or Solions used in the controls of underwater mines. Construction is shown in Fig. 72 as the button type.

The battery's anode is composed of consolidated Grenox or lead–lead oxide. The cathode is composed of consolidated silver powder. When the cell is charged at 1.2 ± 0.03 v, the silver cathode is converted to silver oxide (Ag_2O) and the lead–lead monoxide anode is converted to metallic lead. A Synpor barrier separates the potassium hydroxide electrolyte from the cathode. The electrolyte itself is contained in five thin Dynel absorber pads placed in a pile between the Synpor barrier and the anode. The assembly is pressure sealed with a neoprene grommet which also serves as external insulation between the cathode and anode.

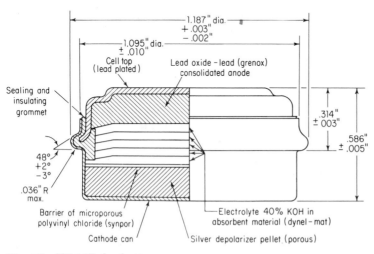

Fig. 72 XZ-10B dry battery.

Sizes and Applications

There have been many attempts at standardization and application. The American Association of Battery Manufacturers has designations, maximum dimensions, and electrical performance under specific test conditions. The Society of Automotive Engineers has a different set of designations. However, the two sets of designations can be correlated as practical equivalents of each other on a performance basis. Many of the manufacturers have different sets of designations, catalog numbers, and dimensions for their product, but a large number of these can be correlated as being operationally equivalent to the AABM and SAE designations. Application tables are presented for special usage of batteries or cells, in radio, instruments, controls, tractors, marine purposes, aircraft of 2, 4, 8, and 24 v (Table 34). Motor cars of 1955 or earlier were on 6-v electrical systems, generators, etc. (Table 35), so that there are correlations in a 6-v table for passenger car units, imported car application, light commercial, golf carts, vehicles, heavy-duty commercial, and the like (Table 35). In the 1950s, car changeover to 12-v electrical systems brought forth 6-cell battery units of 12 v and a number of connector arrangements, cell layouts, and terminals. Table 34 gives the commercial form, dimensions (approximate), correlated designations, electrical performance, application, weight, and the life of the special batteries. Table 35 gives the corresponding data for the 6-v systems, a large number of which are for replacement on older vehicles. Table 36 (12 v) is a correlation of designations of AABM, SAE, ESB, and Delco and relates to original as well as replacement equipment for passenger cars, imported cars, trucks, buses, and heavy-duty applications. In all the tables ampere-hour capacities cover a very wide range. Competition is very keen. No attempt has been made to cover the numerous manufacturers, but Electric Storage Battery, ESB Inc., who claims to be the largest manufacturer of packaged power, and United Delco of General Motors Corporation, one of the largest original-equipment suppliers for automobiles and trucks, are included as representative.

Assemblages for the position of contacts, arrangement of cells, and other data are shown in Fig. 73. Separators are R for rubber, P for plastic, and DB for double insulation which consists of glass retainer mats bonded to rubber separators. For Tables 35 and 36, D is high-polymer dielectric, G is glass, AR is asbestos rubber, F is flat, R and G is rubber and glass fiber, and D and G is dielectric and glass fiber.

Electrical size is often expressed as capacity at the 20-hr rate of discharge at 80°F. If a battery can be discharged at 5 amp for 20 hr before the discharge voltage falls off, the unit is rated at 100 ah.

Suppose 1,200 w be required for 0.1 hr at 0°F. The battery required to deliver this energy can be calculated from the data shown in Fig. 74. For a 0.1-min discharge at 0°F, batteries of the automotive type will

Lead Secondary Cells 127

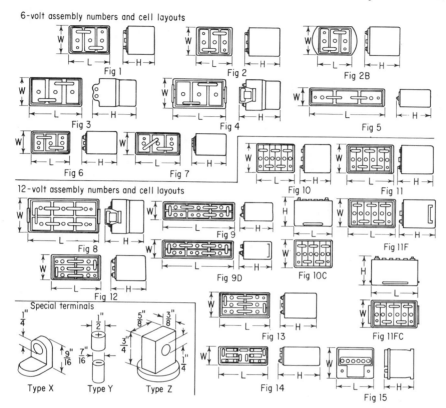

Fig. 73 Assembly numbers of cell layouts, and special terminals.

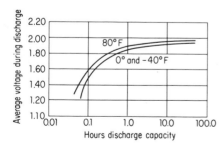

Fig. 74 Average voltage during discharge of automobile batteries.

TABLE 34 2-, 4-, 8-, and 24-volt Lead Storage Batteries

Application	Voltage	Delco	ESB	Terminals*	Equivalent service type	Adjusted units, months	Number of plates per cell	Insulation, type of separator	Pints of acid required	Capacity, 20-ah rating	Maximum dimensions							Approximate weight, lb	
											Length		Width		Height		Dry	Wet
											in.	mm	in.	mm	in.	mm		
Electrolyte retaining	2	...	ER-6-2	B	BB 215/U	...	...	...	...	5.8	1.687	42.84	2.125	53.97	4.00	101.60	0.80	1.01
Electrolyte retaining	2	...	ER-6-2-B	T	...	...	...	...	...	5.8	1.687	42.84	2.375	60.32	4.25	107.95	0.84	1.05
Electrolyte retaining	2	...	ER-11-2-W	W	BB 206/U	...	...	...	...	9.5	2.562	65.07	2.062	52.37	3.906	99.21	1.08	1.45
Electrolyte retaining	2	...	ER-11-2-P	P	...	...	...	...	...	9.5	2.562	65.07	2.062	52.37	4.562	115.87	1.08	1.45
Electrolyte retaining	2	...	Radio 25-2	S	...	...	...	...	...	21.0	2.562	65.07	3.00	76.20	6.00	152.40	2.37	3.00
Electrolyte retaining	2	...	ERH-25-2	B	BB 210/U	...	...	...	...	24.0	2.50	63.50	2.50	63.50	6.50	165.10	2.21	2.95
Electrolyte retaining	2	...	ER-24-2	B	BB 254/U	...	2	...	...	25.0	2.875	73.02	3.843	97.61	5.531	140.48	2.86	3.82
Low discharge type	2	...	DA-2-1	...	...	...	...	...	...	26-30	2.50	63.50	2.50	63.50	6.375	161.92	...	3.5
Electrolyte retaining	2	...	ER-32-2	B	BB 54A	...	...	...	...	32.0	2.875	73.02	3.843	97.61	5.531	140.48	3.20	4.22
Electrolyte retaining	2	...	ER-34-2	P	...	...	...	...	...	34.0	4.50	114.30	3.687	93.64	4.50	114.30	3.65	5.0
Electrolyte retaining	2	...	ER-40-2	B	BB 212/U	...	...	...	...	40.0	6.875	174.62	2.375	60.32	4.50	114.30	4.18	5.30
Low discharge type	2	...	DH-5-1	...	...	...	5	...	...	200	5.50	139.70	7.062	179.37	9.5	241.30	...	26
Low discharge type	2	...	DH-51	...	...	...	5	...	...	500-600	7.562	192.07	8.312	211.12	14.0	355.60	...	58
Electrolyte retaining	4	...	ER-6-4B	T	...	...	...	...	...	5.8	3.375	85.72	2.375	60.32	4.250	107.95	1.75	2.17
Tractor	8	...	TR 8-1	...	...	18	11	D	9	72	9.00	228.60	7.00	177.80	8.750	222.25	29	38
Marine and stationary	8	801	...	...	...	...	15	R	...	105	13.906	353.21	7.125	180.97	9.062	230.17	37	58
Tractor	8	...	TR 8-2	...	...	18	15	D	12	105	10.312	261.92	7.00	177.80	8.750	222.25	36	44
Tractor	8	...	TR 8-3	...	...	18	15	D	13	108	11.625	295.27	7.00	177.80	8.750	222.25	38	50
Tractor	8	...	TR 8-4	...	...	18	17	D	13	115	13.00	330.20	7.00	177.80	9.00	228.60	44	57
Marine and stationary	8	811	...	...	...	...	19	R	...	150	15.968	405.58	7.312	185.72	9.531	242.08	56	76
Marine and stationary	8	821	...	...	...	...	17	DB	...	175	21.437	544.49	7.562	192.07	9.531	242.08	82	103
Aircraft	24	...	24-12	...	...	...	...	...	6	48	7.875	200.02	7.750	196.85	7.375	187.32	30	38
Aircraft	24	...	24-20	...	...	...	...	...	7	96	10.625	269.87	8.00	203.20	8.50	215.90	43	58
Aircraft	24	...	24-35	...	...	...	...	...	10	140	10.375	263.52	10.125	257.17	10.00	254.00	69	82

* Terminals: B—bolt and nut; T—taper posts; W—wire leads; P—7/8 in. pins; SP—3/16 in. pins; F—flat buttons; S—banana plug sockets.

TABLE 35 6-volt Lead Storage Batteries

Application	AABM group 1	SAE	Delco	ESB	Assembly figure	Adjusted units, months	Number of plates per cell
Motorcycle, scooter	.....	.....	...	G-4	.....	...	7
Electrolyte retaining	.....	.....	...	ER-6-6			
Electrolyte retaining	.....	.....	...	ER-6-6B			
Motorcycle, scooter	.....	.....	...	G-2	.....	...	7
Motorcycle, scooter	.....	.....	...	G-3	.....	...	5
Electrolyte retaining	.....	.....	...	ERS-10-6			
Motorcycle, scooter	.....	.....	...	H-3	.....	...	5
Motorcycle, scooter	.....	.....	...	E-2	.....	...	5
Motorcycle, scooter	.....	.....	...	G-1	.....	...	5
Motorcycle, scooter	.....	.....	...	G-3A	.....	...	9
Motorcycle, scooter	.....	.....	...	G-5	.....	...	5
Shipboard gyrocompass	.....	.....	...	NP 15-3R			
Motorcycle, scooter	.....	.....	...	E-1	.....	...	5
Motorcycle, scooter	.....	.....	...	E-1W	.....	...	5
Motorcycle, scooter	.....	.....	...	ID-2	.....	...	5
Electrolyte retaining	.....	.....	...	ER 15-6			
Electrolyte retaining	.....	.....	...	ER 25-6			
Motorcycle, scooter	.....	.....	...	H-2	.....	...	5
Motorcycle, scooter	.....	.....	...	H-1	.....	...	5
Motorcycle, scooter	.....	.....	...	ID-1	.....	...	5
Electrolyte retaining	.....	.....	...	ER 32-6			
Garden equipment	.....	.....	...	HH-1WN	.....	...	5
Electrolyte retaining	.....	.....	...	ER 40-6			
Electrolyte retaining	.....	.....	...	ERS 40-6			
Garden equipment	.....	.....	...	H-4WN	.....	...	7
Imported cars	18LF	.....	...	.....	2B		
Imported cars	18LF						
Imported cars	17HF	.....	...	.....	2B		
Passenger cars	1	1M1	211	.....	.....	...	11
Imported cars	19L	.....	...	.....	2		
Imported cars	17L	.....	...	.....	2		
Imported cars	19L	.....	...	.....	2		
Passenger cars	1	.....	...	13-IR	.....	15	13
Imported cars	18HF	.....	...	.....	2		
Imported cars	18	.....	...	.....	2		
Imported cars	19L	.....	...	.....	2		
Imported cars	19L	.....	...	.....	2		
Imported cars	19	.....	...	.....	2		
Passenger cars	1	.....	...	.....	1		
Passenger cars	1	1M1	311	.....	.....	...	15
Passenger cars	.....	1M1	...				
Passenger cars	.....	.....	...	TR1	.....	24	15
Passenger cars	1	1M1	413	.....	.....	...	13
Passenger cars	2L	.....	...	.....	1		
Passenger cars	2L	3L2	415	.....	.....	...	15
Passenger cars	2E	8H2	...	.....	5		
Passenger cars	.....	1H1	...	.....	2		
Passenger cars	2	1M2	...	.....	2		
Passenger cars	2L	3L2	...	17T	.....	36	17
Passenger cars	.....	3L2					
Passenger cars	2N	.....	...	192N	.....	24	19
Passenger cars	1						
Passenger cars	1	.....	...	15HD	.....	36	15
Passenger cars	1	1M1	513	.....	.....	...	15
Passenger cars	2	1M2	417	.....	.....	...	15
Passenger cars	2E	8H2	419	.....	.....	...	15
Passenger cars	3L	1H3	...	.....	1		
Passenger cars	3N	.....	...	213N	.....	36	21

TABLE 35 6-volt Lead Storage Batteries (Continued)

Application	AABM group 1	SAE	Delco	ESB	Assembly figure	Adjusted units, months	Number of plates per cell
Passenger cars	3NR	....	...	213NL	.....	36	21
Passenger cars	2	....	...	17S-2	.....	36	17
Passenger cars	2F	....	...	17F	.....	24	17
Passenger cars	2E	....	...	17BOP	.....	36	17
Passenger cars	1	1M1	913	....	.....	...	17
Commercial diesel	1	1M1(1)	411	....	.....	...	17
Heavy duty, commercial	1T						
Passenger, light commercial	2	1H2		517	.....	.....	...	17
Passenger, light commercial	2L	3L2					
Passenger, light commercial	2	1H2	...	....	2		
Passenger, light commercial	2E	8H2	...	....	5		
Passenger, light commercial	2F						
Passenger, light commercial	3	....	...	....	2		
Passenger, commercial	1	....	...	19H-1	.....	48	19
Passenger, commercial	2	....	...	2-17T	.....	36	17
Tractors, buses	.....	....	...	COM-3	.....	12	17
Passenger, commercial	2E	....	...	19BOP	.....	48	19
Heavy duty, truck, taxi	.....	....	...	3-17T	.....	18	17
Heavy duty	2	1M2(2)	917	....	.....	...	19
Passenger cars	3	1H3	...	....	2		
Passenger cars	.....	....	...	....	2		
Tractors	3	1H3(3H)	711	....	.....	...	17
Heavy duty, commercial	2T	....	...	....	3 or 4		
Heavy duty, trucks, tractors, buses, taxis	.....	....	...	COM-4	.....	12	19
Heavy duty, trucks, tractors, buses, taxis	2	....	...	21H2	.....	48	21
High duty	3	1H3(3H)	927	....	.....	...	21
Commercial, diesel	3	1H3(311)	713	....	.....	...	19
Passenger cars	4	1H4	...	....	2		
Passenger cars	5	....	...	....	2		
Heavy duty, commercial	3T	6T3	...	....	3 or 4		
High duty	4	1H4(4H)	929	....	.....	...	23
Commercial, diesel	4	1H4(4H)	917	....	.....	...	21
Heavy duty, trucks, tractors, buses, taxis	.....	....	...	COM-4H	.....	18	23
Heavy duty, trucks, tractors, buses, taxis	.....	....	...	5-21T	.....	18	21
Heavy duty, trucks, tractors, buses, taxis	.....	....	...	HD-5D	.....	18	21
Golf carts	.....	....	843	....	.....	...	17
Golf carts	1T	1T4	...	....	.....	...	33
Golf carts	.....	....	847	....	.....	...	17
Heavy duty, trucks, tractors, buses, taxis	.....	....	...	HD-70	.....	18	23
Heavy duty, commercial	5T	....	...	....	3 or 4		
Commercial, diesel	7D	6T3(7D)	719	....	.....	...	27
Heavy duty, commercial	7D	6T3A					
Heavy duty, commercial	9D	....	...	....	2		

TABLE 35 6-volt Lead Storage Batteries (Continued)

Application	Insulation, type of separator	Pints of acid required	Cranking power at 0°F, w	Capacity, 20-ah rating	Electrical values			
					Minutes at 300 amp at 0°F	Minutes at 300 amp, 0°F to 1 volt/cell	Minutes at 300 amp at 0°F, 5 sec/v	Minutes at 300 amp at 0°F, 10 sec/v
Motorcycle, scooter				4.5				
Electrolyte retaining				5.8				
Electrolyte retaining				5.8				
Motorcycle, scooter				7				
Motorcycle, scooter				9				
Electrolyte retaining				10				
Motorcycle, scooter				12				
Motorcycle, scooter				12				
Motorcycle, scooter				12				
Motorcycle, scooter				12				
Motorcycle, scooter				12				
Shipboard gyrocompass				15				
Motorcycle, scooter				18				
Motorcycle, scooter				18				
Motorcycle, scooter				20				
Electrolyte retaining				22				
Electrolyte retaining				24				
Motorcycle, scooter				24				
Motorcycle, scooter				27				
Motorcycle, scooter				27				
Electrolyte retaining				32				
Garden equipment		2		36				
Electrolyte retaining				40				
Electrolyte retaining				40				
Garden equipment		3		48				
Imported cars				50		0.6	3.3	
Imported cars				57		0.85	3.4	
Imported cars				58		0.75	3.4	
Passenger cars	R			65		1.7	3.6	
Imported cars				66		1.2	3.6	
Imported cars				66		1.2	3.6	
Imported cars				70		1.3	3.7	
Passenger cars	D	7		70				
Imported cars				72		1.4	3.7	
Imported cars				75		1.6	3.8	
Imported cars				75		1.5	3.8	
Imported cars				77		1.6	3.8	
Imported cars				84		2.	3.9	
Passenger cars				85		2.5	4.1	
Passenger cars	R			85		3.0	4.1	
Passenger cars				90		3.0	4.1	
Passenger cars	D	7		90				
Passenger cars	R	7		90	2.9		4.2	
Passenger cars				90	2.5		4.1	
Passenger cars	R			90	2.9		4.2	
Passenger cars				100		3.2	4.2	
Passenger cars				100		3.6	4.2	
Passenger cars				100		3.2	4.2	
Passenger cars				100		3.2	4.2	
Passenger cars	D	6		100			4.1	
Passenger cars				100		3.5	4.1	
Passenger cars	D	7		100			4.2	
Passenger cars				105	4.2	3.5		
Passenger cars	D & G	7		105				
Passenger cars	R	7		105	3.7		4.4	
Passenger cars	R			106		3.8		4.4
Passenger cars	R			106		3.8		4.4
Passenger cars				110		4.0	4.3	
Passenger cars	D	7		110				

TABLE 35 6-volt Lead Storage Batteries (Continued)

Application	Insulation, type of separator	Pints of acid required	Cranking power at 0°F, w	Capacity, 20-ah rating	Electrical values			
					Minutes at 300 amp at 0°F	Minutes at 300 amp, 0°F to 1 volt/cell	Minutes at 300 amp, at 0°F, 5 sec/v	Minutes at 300 amp, at 0°F, 10 sec/v
Passenger cars.................	D	7	.....	110				
Passenger cars.................	D & G	7	.....	115				
Passenger cars.................	D	6	.....	115				
Passenger cars.................	D	8	.....	115				
Passenger cars.................	DB	...	.....	115	4.4	.....	....	4.5
Commercial diesel..............	R	...	.....	115	4.1	.....	4.5	
Heavy duty, commercial........		...	.....	118	....	1.5	3.4	
Passenger, light commercial....	F	...	.....	120	....	4.2	4.4	
Passenger, light commercial....		...	.....	120	....	4.5	4.4	
Passenger, light commercial....		...	.....	120	....	4.2	4.4	
Passenger, light commercial....		...	.....	120	....	4.5	4.4	
Passenger, light commercial....		...	.....	120	....	4.5	4.3	
Passenger, light commercial....		...	.....	120	....	4.4	4.3	
Passenger, commercial.........	AR	7	.....	125				
Passenger, commercial.........	R & G	8	.....	125				
Tractors, buses................	D & G	14	.....	125				
Passenger, commercial.........	AR	8	.....	130				
Heavy duty, truck, taxi.........	R & G	9	.....	130				
Heavy duty....................	DB	...	.....	130	5.2	.....	4.6	
Passenger cars.................		...	.....	133	....	5.3	4.4	
Passenger cars.................		...	.....	135	....	5.2	4.4	
Tractors......................	R	...	.....	135	4.8	.....	4.6	
Heavy duty, commercial........		...	.....	137	....	2.7	3.6	
Heavy duty, trucks, tractors, buses, taxis.....................	D & G	15	.....	140				
Heavy duty, trucks, tractors, buses, taxis.....................	AR	6	.....	140				
High duty.....................	DB	...	.....	145	6.0	.....	4.7	
Commercial, diesel.............	R	...	.....	145	6.0	.....	4.7	
Passenger cars.................		...	.....	150	....	6.3	4.5	
Passenger cars.................		...	.....	150	....	6.1	4.5	
Heavy duty, commercial........		...	.....	157	....	4.0	3.8	
High duty.....................	DB	...	.....	160	7.0	.....	4.8	
Commercial, diesel.............	R	...	.....	160	7.0	.....	4.8	
Heavy duty, trucks, tractors, buses, taxis.....................	R	...	.....	160				
Heavy duty, trucks, tractors, buses, taxis.....................	R & G	11	.....	165				
Heavy duty, trucks, tractors, buses, taxis.....................	R & G	11	.....	165				
Golf carts.....................	DB	...	.....	165				
Golf carts.....................	R	...	4,500	175	....	11.5	5.0	
Golf carts.....................	DB	...	.....	180				
Heavy duty, trucks, tractors, buses, taxis.....................	R & G	12	.....	185				
Heavy duty, commercial........		...	.....	196	....	6.4	4.0	
Commercial, diesel.............	R	...	.....	200	10.5	.....	4.9(30 sec/v)	
Heavy duty, commercial........		...	.....	200	....	10.5	....	4.9(30 sec/v)
Heavy duty, commercial........		...	.....	336	....	17.4	....	5.1(30 sec/v)

TABLE 35 6-volt Lead Storage Batteries (Continued)

Application	Length in.	Length mm	Width in.	Width mm	Height in.	Height mm	Weight Dry	Weight Wet	Current acceptance	Overcharge cycles
Motorcycle, scooter	5.000	127.00	2.062	52.37	4.186	106.32	3			
Electrolyte retaining	5.063	128.60	2.125	53.97	4.0	101.60	2.44	3.09		
Electrolyte retaining	5.063	128.60	2.375	60.32	4.25	107.95	2.65	3.28		
Motorcycle, scooter	5.000	127.00	2.062	52.37	5.00	127.00	3			
Motorcycle, scooter	4.806	122.07	2.375	60.32	5.375	136.52	4			
Electrolyte retaining	6.063	154.00	2.531	64.28	4.75	120.65	3.80	4.75		
Motorcycle, scooter	3.625	92.07	4.250	107.95	5.750	146.05	7	8		
Motorcycle, scooter	3.625	92.07	3.250	82.55	5.875	149.22	5	7		
Motorcycle, scooter	3.625	92.07	3.250	82.55	6.500	165.10	5	7		
Motorcycle, scooter	4.806	122.07	2.375	60.32	5.375	136.52	5			
Motorcycle, scooter	9.186	233.32	1.434	36.42	6.250	158.75	6			
Shipboard gyrocompass	4.50	114.30	4.625	117.47	6.50	165.10	10.50	12.50		
Motorcycle, scooter	3.625	92.07	4.750	120.65	6.625	168.27	8	10		
Motorcycle, scooter	3.625	92.07	4.750	120.65	6.625	168.27	8	10		
Motorcycle, scooter	4.434	112.62	3.500	88.90	7.125	180.97	8	10		
Electrolyte retaining	8.375	212.72	3.840	97.53	4.375	111.12	7.75	9.75		
Electrolyte retaining	7.500	190.50	2.50	63.50	6.50	165.10	6.88	9.10		
Motorcycle, scooter	4.500	114.30	3.875	98.42	8.625	219.07	10	13		
Motorcycle, scooter	4.000	101.60	6.500	165.10	6.625	168.27	11	14		
Motorcycle, scooter	3.875	98.42	6.500	165.10	7.500	190.50	11	14		
Electrolyte retaining	8.625	219.07	3.840	97.53	5.531	140.48	9.85	12.91		
Garden equipment	3.875	98.42	6.500	165.10	8.000	203.20	13	15		
Electrolyte retaining	7.125	180.97	6.875	174.62	4.806	122.07	12.79	16.15		
Electrolyte retaining	8.806	223.67	3.930	99.82	6.806	172.87	12	16		
Garden equipment	4.559	115.79	6.500	165.10	8.750	222.25	19	22		
Imported cars	8.437	214.29	6.875	174.62	7.500	190.50			4.0	1
Imported cars	8.437	214.29	6.875	174.62	7.500	190.50			4.3	2
Imported cars	7.375	187.32	6.875	175.62	9.000	228.60			4.4	3
Passenger cars	8.968	227.78	7.000	177.80	8.656	219.86	20	22		
Imported cars	8.500	215.90	7.000	177.80	7.500	190.50			5	3
Imported cars	7.375	187.32	6.620	168.14	7.55	191.77			5.0	3
Imported cars	8.50	215.90	7.00	177.80	7.50	190.50			5.3	3
Passenger cars	9.00	228.60	7.00	177.80	8.75	222.25		30		
Imported cars	8.437	214.29	6.875	174.62	9.250	234.95			5.4	2
Imported cars	7.750	196.85	6.875	174.62	8.50	215.90			5.6	4
Imported cars	8.50	215.90	7.00	177.80	7.50	190.50			5.6	4
Imported cars	8.50	215.90	7.00	177.80	7.50	190.50			5.7	4
Imported cars	8.937	226.99	6.812	172.99	8.50	215.90			6.3	4
Passenger cars	9.125	231.77	7.125	180.97	9.375	238.12			6.4	5
Passenger cars	8.968	227.87	7.000	177.80	8.656	219.86	25	36		
Passenger cars	9.125	231.77	7.125	180.97	9.187	233.34			6.8	5
Passenger cars	9.000	228.60	7.000	177.80	8.750	222.25	31	39		
Passenger cars	9.187	233.34	7.062	179.37	8.656	219.86	25	36		
Passenger cars	10.620	269.74	7.250	184.15	8.125	206.37			6.8	5
Passenger cars	10.625	269.87	7.06	179.32	7.906	200.81	25	37		
Passenger cars	19.375	492.12	4.125	104.77	9.125	231.77			7.5	6
Passenger cars	9.125	231.77	7.125	180.97	9.375	238.12			7.5	6
Passenger cars	9.125	231.77	7.125	180.97	9.187	233.34			7.5	5
Passenger cars	10.372	263.44	7.125	180.97	9.375	238.12			7.5	6
Passenger cars	10.625	269.87	7.125	180.97	7.875	200.02	30	40	7.5	6
Passenger cars	10.625	269.87	7.250	184.15	8.125	206.37			7.5	6
Passenger cars	10.000	254.00	5.562	141.27	8.937	226.99	28	38	7.5	6
Passenger cars	9.125	231.77	7.125	180.97	9.187	233.34			7.2	7
Passenger cars	9.000	228.60	7.00	177.80	8.750	222.25	28	39		
Passenger cars	9.187	233.34	7.125	180.97	7.875	200.02	27	37		
Passenger cars	10.372	263.44	7.062	179.37	8.656	219.86	29	41		
Passenger cars	19.372	492.04	4.062	103.17	8.812	223.82	30	43		
Passenger cars	11.875	301.62	7.250	184.15	8.125	206.37			8.3	7
Passenger cars	11.250	285.75	5.50	139.70	8.875	225.42	32	42		

TABLE 35 6-volt Lead Storage Batteries (Continued)

Application	Length in.	Length mm	Width in.	Width mm	Height in.	Height mm	Approximate weight, lb Dry	Approximate weight, lb Wet	Current acceptance	Overcharge cycles
Passenger cars	11.250	285.75	5.50	139.70	8.875	225.42	32	42		
Passenger cars	10.312	261.92	7.00	177.80	8.750	222.25	35	43		
Passenger cars	15.500	266.70	7.250	184.15	9.125	231.77	38	47		
Passenger cars	19.250	488.95	4.00	101.60	8.750	222.25	34	47		
Passenger cars	8.967	277.61	7.031	178.58	9.062	230.17	31	40		
Commercial diesel	8.967	277.61	6.937	176.19	9.062	230.17	29	40		
Heavy duty, commercial	13.375	339.72	7.625	193.67	10.50	266.70	.....	.....	8.9	8
Passenger, light commercial	10.375	263.52	7.125	180.97	9.375	238.12	31	43	9.0	8
Passenger, light commercial	10.625	269.87	7.125	180.97	8.125	206.37	.....	.....	9.0	8
Passenger, light commercial	10.375	263.52	7.125	180.97	9.375	238.12	.....	.....	9.0	8
Passenger, light commercial	19.375	492.12	4.125	104.77	9.125	231.77	.....	.....	9.0	8
Passenger, light commercial	10.625	269.87	7.312	185.72	9.375	238.12	.....	.....	9.0	8
Passenger, light commercial	11.750	298.45	7.125	180.97	9.375	238.12	.....	.....	9.0	8
Passenger, commercial	9.00	228.60	7.00	177.80	8.750	222.25	33	43		
Passenger, commercial	10.312	261.92	7.00	177.80	8.750	222.25	35	51		
Tractors, buses	11.687	296.84	7.062	179.37	9.280	235.71	32	50		
Passenger, commercial	19.250	488.95	4.00	101.60	8.750	222.25	36	49		
Heavy duty, truck, taxi	11.625	295.27	7.125	180.97	9.062	230.17	43	57		
Heavy duty	10.375	263.52	7.031	178.58	9.062	230.17	35	47		
Passenger cars	11.750	298.45	7.125	180.97	9.375	238.12	.....	.....	10.0	9
Passenger cars	13.125	333.37	7.125	180.97	9.375	238.12	.....	.....	10.0	10
Tractors	11.625	295.27	7.062	179.37	9.062	230.17	37	50		
Heavy duty, commercial	15.125	384.17	7.625	193.67	10.50	266.70	.....	.....	10.3	10
Heavy duty, trucks, tractors, buses, taxis	12.875	327.02	7.062	179.37	9.280	235.71	38	54		
Heavy duty, trucks, tractors, buses, taxis	10.312	261.92	7.00	177.80	8.750	222.25	36	48		
High duty	11.625	295.27	7.062	179.37	9.062	230.17	42	57		
Commercial, diesel	11.625	295.27	7.062	179.37	9.062	230.17	41	55		
Passenger cars	13.125	333.37	7.125	180.97	9.375	238.12	.....	.....	10.9	11
Passenger cars	14.50	368.30	7.125	180.97	9.125	231.77	.....	.....	11.3	11
Heavy duty, commercial	16.875	428.62	7.625	193.67	10.50	266.70	.....	.....	11.8	12
High duty	13.062	331.77	7.062	179.37	9.062	230.17	46	61		
Commercial, diesel	13.062	331.77	7.062	179.37	9.062	230.17	44	59		
Heavy duty, trucks, tractors, buses, taxis	12.875	327.02	7.062	179.37	9.280	235.71	38	56		
Heavy duty, trucks, tractors, buses, taxis	14.50	368.30	7.375	187.32	8.750	222.25	53	72		
Heavy duty, trucks, tractors, buses, taxis	13.50	342.90	7.00	177.80	9.50	241.30	52	70		
Golf carts	10.250	260.35	7.062	179.37	10.280	261.11	49	63		
Golf carts	13.00	330.20	7.00	177.80	9.50	241.30	53	69		
Golf carts	11.625	295.27	7.062	179.37	10.812	274.62	54	67		
Heavy duty, trucks, tractors, buses, taxis	16.50	421.64	7.50	190.50	9.50	241.30	59	83		
Heavy duty, commercial	20.875	530.22	7.625	193.67	10.50	266.70	.....	.....	14.7	16
Commercial, diesel	15.937	404.79	7.093	180.16	9.031	229.38	55	73		
Heavy duty, commercial	16.250	412.75	7.125	180.97	9.375	238.12	.....	.....	15.0	16
Heavy duty, commercial	25.50	647.70	7.50	190.50	11.562	293.67	.....	.....	25.2	30

Lead Secondary Cells

TABLE 36 12-volt Lead Storage Batteries

Application	AABM group 1	SAE	Delco	ESB	Assembly figure	Type	Adjusted units, months	Number of plates per cell	Insulation, type of separator
Imported cars	21NL	.....	...	.....	11◆(Z)				
Passenger cars	53	14M2	...	.....	14				
Imported cars	22NL	.....	...	.....	11◆(Y)				
Passenger cars	20H	9HCO	...	.....	10				
Passenger cars	22NF	18M1	...	.....	11F				
Imported cars	21SL	.....	...	.....	11◆(Z)				
Passenger cars	24	9M3	...	.....	10				
Passenger cars	20H	.....	...	.....	.....	HD20H	36	7	D
Cranking power	24F	17M2	G49	.....	.....	.....	...	7	P
Cranking power	22FC	17MJ1	G55	.....	.....	.....	...	7	P
Cranking power	24C	9MJ3	G59	.....	.....	.....	...	7	P
Cranking power	29NF	4NF	G65	.....	.....	.....	...	9	P
Inboard/outboard marine	....	.....	...	2SM/24	.....	7SM-2W	12	7	D
Inboard/outboard marine	....	.....	...	2SM/24	.....	7SM-2W(2)	12	7	D
Passenger cars	53	14M2	557	.....	.....	.....	...	9	R
Passenger cars	22F	17M1	...	.....	11F				
Passenger cars	22NF	.....	...	.....	.....	HD22NF	36	9	D
Passenger cars	53	.....	...	.....	.....	HD53KM	36	7	D
Passenger cars	22F	.....	...	.....	.....	HD215M	36	7	D
Passenger cars	22FC	17MJ1B	...	.....	11FC				
Cranking power	22FC	17MJ1	Y55	.....	.....	.....	...	9	R
Imported cars	....	.....	...	VW12	.....	.....	36	11	
Passenger cars	22HF	17M1A	...	.....	11F				
Passenger cars	24	9M3A	...	.....	.....				
Passenger cars	24	.....	...	2SM	.....	FL2SM	24	7	D
Passenger cars	60	.....	...	3KM	.....	FL3KM	24	7	D
Passenger cars	24H	9HCB	...	.....	10				
Passenger cars	24	9M3A	...	.....	10				
Cranking power	60	3KM	G63	.....	.....	.....	...	9	P
Passenger cars	24	9M3B	...	.....	10				
Imported cars	24H	9H3C	...	.....	10	.....	36		
Passenger cars	24H	9H3	...	.....	10				
Passenger cars	24	9M3B	...	.....	10				
Cranking power	24F	17M2	Y49	.....	.....	.....	...	9	R
Cranking power	24C	9MJ3	Y59	.....	.....	.....	...	9	R
Passenger cars	24C	9MJ3C	...	.....	10C				
Passenger cars	24F	9MJ3F	...	.....	11F				
Passenger-commercial	29NF	.....	...	4NF	.....	HD4NF	36	11	D
Cranking power	29NF	4NF	Y65	.....	.....	.....	...	11	R
Cranking power	60	3KM	Y63	.....	.....	.....	...	11	R
Imported cars	28SM	.....	...	.....	11				
Imported cars	29H	.....	...	.....	10				
Passenger cars	24H	9HC3A	...	.....	10				
Inboard/outboard marine	....	.....	...	2SM/24	.....	9SM-2W	24	9	D & G
Passenger cars	24	9M3D	...	.....	10				
Passenger cars	60	15M4	...	.....	12				
Passenger cars	24	.....	...	2SM	.....	HD2SM	36	9	D & G
Passenger cars	24F	.....	...	24F	.....	HD24F	36	9	D & G
Passenger cars	60	.....	...	3KM	.....	HD3KM	36	9	D & G
Passenger cars	24H	9H3A	...	.....	10				
Passenger cars	24	9M3D	...	.....	10				
Inboard/outboard marine	....	.....	...	2SM/24	.....	9SM-2W(2)	24	9	D & G
Cranking power	24C	9MJ3	R59	.....	.....	.....	...	11	R
Passenger cars	24C	9MJ3C	...	.....	10C				
Passenger cars	60	15M4	...	.....	.....				
Marine and stationary	....	.....	859	.....	.....	.....	...	11	R
Cranking power	29NF	4NF	R65	.....	.....	.....	...	13	R
Passenger cars	24F	.....	...	.....	.....				
Passenger cars	29NF	18M3A	...	.....	.....				
Passenger cars	32N	.....	...	.....	11				
Passenger cars	32N	.....	...	5SN	.....	HD5SN	36	13	D
Passenger cars	24	.....	...	2SM	.....	SU2SM	48	9	AR

TABLE 36 12-volt Lead Storage Batteries (Continued)

Application	AABM group 1	SAE	Delco	ESB	Assembly figure	Type	Adjusted units, months	Number of plates per cell	Insulation, type of separator
High duty	29NF	18M3(4NF)	965	......	......	....	...	13	DB
High duty	24	9M3(2SM)	959	......	......	....	...	11	DB
Passenger cars	24H	9H3B	...	......	10				
Passenger cars	29NF	....	...	......	......	SU4NF	48	13	AR
Cranking power	60	3KM	R63	......	......	....	...	11	R
Cranking power	27C	9MJ6	Y71	......	......	....	...	11	R
Cranking power	27F	3MFA	Y81	......	......	....	...	11	R
Passenger cars	24H	9HC3B	...	......	10				
Passenger cars	24T	9T3	...	......	10				
Passenger cars	27	....	...	3SM	......	HD3SM	36	11	D & G
Passenger cars	27F	....	...	3MS	......	HD3MS	36	11	D & G
Passenger cars	24H	....	...	2SM	......	SU24H	48	11	AR
Cranking power	24TC	9TJ3	...	......	10C	....	...	11	R
Passenger cars	27	9M6	...	......	10				
Passenger cars	27F	17M3B	...	......	11F				
Passenger cars	27H	9H5	...	......	10				
Passenger cars	27H	9HC5	...	......	10				
Passenger cars	27HF	17H3	...	......	11F				
Passenger cars	60	15M4A	...	......	12				
Passenger cars	60	....	...	3KM	......	SU3KM	48	11	AR
Passenger cars	27	....	...	3SM	......	SU3SM	48	11	AR
Passenger cars	32N	11M6A	...	......	11				
Inboard/outboard marine	.....	....	...	2SM/24	......	SM72(1)	24	11	D
Inboard/outboard marine	.....	....	...	3SM/27	......	11SM3W(2)	24	11	D & G
Cranking power	27C	9MJ6	R71	......	10C	....	...	13	R
Passenger cars	27F	17M3							
Tractors	3ET	13TC2	361	......	......	....	...	13	R
High duty	27	9H3(3SM)	971	......	......	....	...	13	DB
Passenger cars	27HF	17H3	...	......	11F				
Taxis	30HR	....	...	C30-HR	......	....	12	9	D & G
Tractors	3ET	....	...	HD-3ET	......	....	18	11	D
Passenger cars	29HR	17H3A	...	......	11				
High duty	30H	9H9	975	......	......	....	...	15	DB
Inboard/outboard marine	.....	....	...	5SH/30	......	13SM-SW	24	13	AR
Heavy duty	5SH	....	...	C-5SH	18	....	...	13	AR
Inboard/outboard marine	.....	....	...	5SH/30	......	13SM-5W(2)	24	13	AR
Ordnance	.....	....	837	......	......	6TN	...	23	P
Tractors	.....	....	...	TR12-95	......	....	18	11	AR
Lift trucks	.....	....	...	BJ-100	......	....	12	13	D & G
Heavy duty	1B	....	...	17-6-3	......	....	24	17	R & G
Tractors	.....	....	...	TR12-135	......	....	18	15	AR
Heavy duty	2B	....	...	19-6-3	......	....	24	19	R & G
Diesel	4D	20T4(4D)	759	......	......	....	...	19	R
High duty	HD	20T4(4D)	983	......	......	....	...	21	DB
Heavy duty	3B	....	...	21-6-3	......	....	24	21	R & G
High duty	4B	20T8(8G)	987	......	......	....	...	17	DB
Heavy duty	4B	....	...	23-6-3	......	....	24	23	R & G
Heavy duty	HD	....	...	27-6-3	......	....	24	27	AR
Ordnance	.....	....	839	......	......	8T	...	37	P
High duty	8D	20T8(8D)	985	......	......	....	...	29	DB
Diesel	8D	20T8(8D)	761	......	......	....	...	27	R

TABLE 36 12-volt Lead Storage Batteries (Continued)

				Electrical values					
Application	Pints of acid required	Cranking power at 0°F, w	Capacity, 20-ah rating	300 amp at 0°F, min	300 amp at 0°F, 10 sec/v	Minutes at 0°F, 150 amp to 1 volt/cell	150 amp at 0°F, 5 sec/v	Amp for 1.5 min	
								0°F	32°F
Imported cars	...	.....	24	....	....	1.3	7.3		
Passenger cars	...	.....	35						
Imported cars	...	.....	38	....	....	2.2	8.1		
Passenger cars	...	.....	38	....	....	2.5	8.4		
Passenger cars	...	.....	40						
Imported cars	...	.....	40	....	....	2.75	8.4		
Passenger cars	...	.....	40	....	....	2.3	8.2		
Passenger cars	7	.....	40						
Cranking power	...	1,800	41	0.6	6.3	2.3	8.3		
Cranking power	...	1,800	41	0.6	6.3	2.3	8.3		
Cranking power	...	1,800	41	0.6	6.3	2.3	8.3		
Cranking power	...	2,000	41	0.7	6.3	2.6	8.4		
Inboard/outboard marine	11	.....	41						
Inboard/outboard marine	11	.....	41						
Passenger cars	...	.....	42	....	....	3.1	8.4		
Passenger cars	...	.....	42	....	....	3.1	8.4		
Passenger cars	6	.....	42						
Passenger cars	7	.....	42						
Passenger cars	7	.....	42						
Passenger cars	...	.....	44	....	....	3.1	8.4		
Cranking power	...	2,300	45	1.0	6.6	3.2	8.5		
Imported cars	10	.....	45						
Passenger cars	...	.....	45	....	....	3.5	9.3		
Passenger cars	...	.....	45	....	....	2.5	8.4		
Passenger cars	9	.....	47						
Passenger cars	11	.....	47						
Passenger cars	...	.....	48	....	....	3.6	8.5		
Passenger cars	...	.....	48	....	....	3.6	8.5		
Cranking power	...	2,400	50	1.2	7.0	3.7	9.2		
Passenger cars	...	.....	50	....	....	3.7	8.4		
Imported cars	7	.....	50						
Passenger cars	...	.....	50	....	....	3.9	8.6		
Passenger cars	...	.....	50	....	....	3.7	8.4		
Cranking power	...	2,350	53	1.1	6.8	4.0	9.0		
Cranking power	...	2,350	53	1.1	6.8	4.0	9.0		
Passenger cars	...	.....	53	....	....	3.8	8.9		
Passenger cars	...	.....	55	....	....	4.4	8.6		
Passenger-commercial	10	.....	55						
Cranking power	...	2,450	55	1.3	7.2	3.8	9.4		
Cranking power	...	2,700	56	1.4	7.5	4.8	9.4		
Imported cars	...	.....	56	....	....	4.1	9.0		
Imported cars	...	.....	57	....	....	4.9	9.1		
Passenger cars	...	.....	59	....	....	5.2	9.2		
Inboard/outboard marine	9	.....	60						
Passenger cars	...	.....	60	....	....	4.4	9.2		
Passenger cars	...	.....	60	....	....	4.4	8.6		
Passenger cars	9	.....	60						
Passenger cars	9	.....	60						
Passenger cars	11	.....	60						
Passenger cars	...	.....	60	....	....	5.5	9.2		
Passenger cars	...	.....	60	....	....	4.4	9.2		
Inboard/outboard marine	9	.....	60						
Cranking power	...	2,900	61	1.6	7.9	5.6	9.5		
Passenger cars	...	.....	61	....	....	5.6	9.4		
Passenger cars	...	.....	62	....	....	4.4	9.2		
Marine and stationary	...	.....	63						
Cranking power	...	2,750	65	2.0	7.7	5.8	9.6		
Passenger cars	...	.....	65	....	....	5.7	9.3		

TABLE 36 12-volt Lead Storage Batteries (Continued)

Application	Pints of acid required	Cranking power at 0°F, w	Capacity, 20-ah rating	Electrical values				Amp for 1.5 min	
				300 amp at 0°F, min	300 amp at 0°F, 10 sec/v	Minutes at 0°F, 150 amp to 1 volt/cell	150 amp at 0°F, 5 sec/v	0°F	32°F
Passenger cars	...	.....	65	....	....	5.8	9.4		
Passenger cars	...	.....	65	....	....	5.0	9.0		
Passenger cars	11	.....	65						
Passenger cars	9	.....	65						
High duty	...	.....	65	2.0	7.5				
High duty	...	.....	67	2.1	7.5				
Passenger cars	...	.....	67	....	....	5.7	9.3		
Passenger cars	10	.....	67						
Cranking power	...	3,000	70	2.0	7.8	5.8	9.5		
Cranking power	...	3,000	70	2.0	7.8	5.8	9.5		
Cranking power	...	3,000	70	2.0	7.8	5.8	9.5		
Passenger cars	...	.....	70	....	....	6.0	9.3		
Passenger cars	...	.....	70	....	....	6.0	9.3		
Passenger cars	10	.....	70						
Passenger cars	10	.....	70						
Passenger cars	9	.....	70						
Cranking power	...	3,150	70	2.1	7.9	6.0	9.8		
Passenger cars	...	.....	70	....	....	5.8	9.0		
Passenger cars	...	.....	70	....	....	5.8	9.0		
Passenger cars	...	.....	70	....	....	7.2	9.8		
Passenger cars	...	.....	70	....	....	6.4	9.6		
Passenger cars	...	.....	70	....	....	5.8	9.6		
Passenger cars	...	.....	70	....	....	5.8	9.4		
Passenger cars	11	.....	72						
Passenger cars	9	.....	72						
Passenger cars	...	.....	72	....	....	7.0	9.8		
Inboard/outboard marine	...	.....	72						
Inboard/outboard marine	10	.....	72						
Cranking power	...	3,350	73	2.1	8.5	6.2	9.9		
Passenger cars	...	.....	75	....	....	7.0	9.8		
Tractors	...	.....	75	2.1	8.2				
High duty	...	.....	75	2.3	7.8				
Passenger cars	...	.....	80	....	....	5.8	9.6		
Taxis	13	.....	80						
Tractors	12	.....	80						
Passenger cars	...	.....	85	....	....	7.8	10.1		
High duty	...	.....	85	2.8	8.0				
Inboard/outboard marine	13	.....	90						
Heavy duty	12	.....	90						
Inboard/outboard marine	13	.....	90						
Ordnance	...	.....	100	1.25(−40°F)	7.2(−40°F; 5 sec/v)				
Tractors	12	.....	100						
Lift trucks	14	.....	100						
Heavy duty	...	.....	120						
Tractors	18	.....	135						
Heavy duty	...	.....	145						
Diesel	...	.....	150	6.0	9.3(30 sec/v)	....	....	700	850
High duty	...	.....	150	6.0					
Heavy duty	...	.....	165						
High duty	...	.....	175	5.0	8.8				
Heavy duty	...	.....	185						
Heavy duty	...	.....	190						
Ordnance	...	.....	200	4.0(at −40°F)	8.8 (at −40°F; 5 sec/v)				
High duty	...	.....	205	10.5	9.8				
Diesel	...	.....	205	10.5	9.8(30 sec/v)	....	....	950	1,100

TABLE 36 12-volt Lead Storage Batteries (Continued)

Application	Length in.	Length mm	Width in.	Width mm	Height in.	Height mm	Approximate weight, lb Dry	Approximate weight, lb Wet	Current acceptance	Overcharge cycles
Imported cars	9.000	228.60	4.875	123.82	6.500	165.10	...	...	....	2
Passenger cars	13.000	330.20	4.682	118.92	8.250	209.55	...	...	4.0	4
Imported cars	9.250	234.95	5.250	133.35	7.812	198.42	...	...	4.0	4
Passenger cars	7.812	198.42	6.750	171.45	9.312	236.52	...	...	4.0	4
Passenger cars	9.437	239.69	5.500	139.70	8.937	226.99	...	...	4.0	4
Imported cars	9.000	228.60	6.687	169.84	8.000	203.20	...	...	4.0	4
Passenger cars	10.250	260.35	6.812	173.02	8.875	225.42	...	...	4.0	4
Passenger cars	7.75	196.85	6.750	171.45	9.312	236.52	25	34		
Cranking power	10.250	260.35	6.812	173.02	8.750	222.25	25	38		
Cranking power	9.500	241.30	6.812	173.02	8.312	211.12	22	32		
Cranking power	10.250	260.35	6.812	173.02	8.312	211.12	25	38		
Cranking power	13.000	330.20	5.437	138.09	8.812	223.82	26	38		
Inboard/outboard marine	13.250	260.35	6.875	174.62	9.125	231.77	31	39		
Inboard/outboard marine	10.250	260.35	6.875	174.62	9.125	231.77	31	39		
Passenger cars	13.062	331.77	4.781	121.43	8.312	211.12	26	35		
Passenger cars	9.500	241.30	6.875	174.62	8.625	219.07	...	...	4.0	4
Passenger cars	9.500	241.30	5.500	139.70	9.000	228.60	27	35		
Passenger cars	13.000	330.20	4.750	120.65	8.437	214.29	28	34		
Passenger cars	8.937	226.99	6.875	174.62	8.312	211.12	27	34		
Passenger cars	9.500	241.30	6.875	174.62	8.312	211.12	...	...	4.0	4
Cranking power	9.500	241.30	6.812	173.02	8.250	209.55	25	34		
Imported cars	9.500	241.30	6.875	174.62	6.625	168.27	26	37		
Passenger cars	9.500	241.30	6.875	174.62	6.875	174.62	...	...	4.0	5
Passenger cars	10.250	260.35	6.812	173.02	8.875	225.42	...	...	4.0	5
Passenger cars	10.250	260.35	6.875	174.62	9.000	228.60	...	41		
Passenger cars	13.000	330.20	6.250	158.75	8.875	225.42	...	45		
Passenger cars	10.250	260.35	6.812	173.02	9.375	238.12	...	...	4.0	5
Passenger cars	10.250	260.35	6.812	173.02	8.875	225.42	...	...	4.0	5
Cranking power	13.000	330.20	6.250	158.75	8.750	222.25	30	45		
Passenger cars	10.250	260.35	6.812	173.02	8.875	225.42	...	...	4.0	5
Imported cars	11.625	295.27	6.750	171.45	7.062	179.37	30	38		
Passenger cars	10.250	260.35	6.812	173.02	9.375	238.12	...	...	4.0	5
Passenger cars	10.250	260.35	6.812	173.02	8.875	225.42	...	...	4.0	5
Cranking power	10.750	273.05	6.812	173.02	8.750	222.25	29	41		
Cranking power	10.250	260.35	6.812	173.02	8.750	222.25	29	41		
Passenger cars	10.250	260.35	6.812	173.02	8.875	225.42	...	...	4.0	5
Passenger cars	10.250	260.35	6.812	173.02	9.000	228.60	...	...	4.1	6
Passenger-commercial	12.500	317.50	5.500	139.70	8.937	226.99	34	41		
Cranking power	13.000	330.20	5.437	138.09	8.750	222.25	31	41		
Cranking power	13.000	330.20	6.250	158.75	8.750	222.25	34	47		
Imported cars	12.187	309.54	7.687	195.24	8.625	219.07	...	...	4.2	6
Imported cars	12.750	323.85	6.875	174.62	9.062	230.17	...	...	4.2	6
Passenger cars	10.250	260.35	6.812	173.02	9.375	238.12	...	...	4.5	6
Inboard/outboard marine	10.250	260.35	6.875	174.62	9.125	231.77	35	42		
Passenger cars	10.250	260.35	6.812	173.02	8.875	225.42	...	...	4.5	6
Passenger cars	13.062	331.77	6.312	160.32	8.875	225.42	...	...	4.5	6
Passenger cars	10.250	260.35	6.875	174.62	9.000	228.60	33	43		
Passenger cars	10.250	260.35	6.875	174.62	9.000	228.60	33	43		
Passenger cars	13.000	330.20	6.250	158.75	8.875	225.42	34	50		
Passenger cars	10.250	260.35	6.812	173.02	9.375	238.12	...	...	4.5	6
Passenger cars	10.250	260.35	6.812	173.02	8.875	225.42	...	...	4.5	6
Inboard/outboard marine	10.250	260.35	6.875	174.62	9.125	231.77	35	42		
Cranking power	10.250	260.35	6.812	173.02	8.750	222.25	33	44		
Passenger cars	10.250	260.35	6.812	173.02	8.875	225.42	...	...	4.5	6
Passenger cars	13.062	331.77	6.312	160.32	8.875	225.42	...	...	4.5	6
Marine and stationary	10.750	273.05	6.812	173.02	9.125	231.77	35	46		
Cranking power	13.000	330.20	5.437	138.09	8.750	222.25	35	45		
Passenger cars	10.750	273.05	6.812	173.02	9.000	228.60	...	...	4.9	7

TABLE 36 12-volt Lead Storage Batteries (Continued)

	Maximum dimensions						Approximate weight, lb		Current acceptance	Overcharge cycles
	Length		Width		Height					
Application	in.	mm	in.	mm	in.	mm	Dry	Wet		
Passenger cars	13.000	330.20	5.500	139.70	8.937	226.99	...	...	4.87	7
Passenger cars	14.250	361.95	5.500	139.70	8.937	226.99	...	...	4.87	7
Passenger cars	14.250	361.95	5.500	139.70	8.937	226.99	38	53		
Passenger cars	10.250	260.35	6.875	174.62	9.000	228.60	35	42		
High duty	12.937	328.59	5.437	138.09	8.906	226.21	38	47		
High duty	10.250	260.35	6.812	173.02	9.407	238.93	40	50		
Passenger cars	10.250	260.35	6.812	173.02	9.375	238.12	...	...	5.0	7
Passenger cars	12.250	311.15	5.500	139.70	8.937	226.99	34	41		
Cranking power	13.000	330.20	6.250	158.75	8.750	222.25	38	51		
Cranking power	12.031	305.58	6.812	173.02	8.750	222.25	37	50		
Cranking power	12.500	317.50	6.812	173.02	8.750	222.25	37	50		
Passenger cars	10.250	260.35	6.812	173.02	9.375	238.12	...	...	5.25	7
Passenger cars	10.250	260.35	6.812	173.02	9.875	250.82	...	...	5.25	7
Passenger cars	11.750	298.45	6.875	174.62	9.125	231.77	39	53		
Passenger cars	11.750	298.45	6.875	174.62	9.125	231.77	39	53		
Passenger cars	10.250	260.35	6.875	174.62	9.312	236.52	37	45		
Cranking power	10.250	260.35	6.812	173.02	9.500	241.30	37	49		
Passenger cars	12.062	306.37	6.407	162.73	8.875	225.42	...	...	5.25	7
Passenger cars	12.500	317.50	6.407	162.73	8.937	226.99	...	...	5.25	7
Passenger cars	11.750	298.45	6.812	173.02	9.250	234.95	...	...	5.25	7
Passenger cars	11.750	298.45	6.812	173.02	9.250	234.95	...	...	5.25	7
Passenger cars	12.500	317.50	6.812	173.02	6.407	162.73	...	...	5.25	7
Passenger cars	13.062	331.77	6.312	160.32	8.875	225.42	...	...	5.25	7
Passenger cars	13.000	330.20	6.250	158.75	8.875	225.42	42	56		
Passenger cars	11.275	286.38	6.875	174.62	9.125	231.77	42	54		
Passenger cars	14.125	358.77	5.500	139.70	8.469	215.11	...	...	5.4	7
Inboard/outboard marine	10.250	260.35	6.875	174.62	9.125	231.77	33	46		
Inboard/outboard marine	11.750	298.45	6.875	174.62	9.125	231.77	42	53		
Cranking power	12.062	306.37	6.812	173.02	8.750	222.25	38	51		
Passenger cars	12.500	317.50	6.407	162.73	8.937	226.99	...	...	5.6	8
Tractors	19.312	490.52	4.344	110.33	9.781	250.72	41	56		
High duty	12.062	306.37	6.812	173.02	9.407	238.93	46	57		
Passenger cars	12.500	317.50	6.812	173.02	9.187	233.34	...	...	5.25	7
Taxis	13.500	342.90	6.812	173.02	9.219	234.16	49	65		
Tractors	19.125	485.77	4.250	107.95	9.250	247.65	48	63		
Passenger cars	13.125	333.75	6.812	173.02	9.125	231.77	...	...	8.5	18
High duty	13.500	342.90	6.812	173.02	9.125	231.77	48	63		
Inboard/outboard marine	13.500	342.90	6.875	174.62	9.125	231.77	54	68		
Heavy duty	13.500	342.90	6.875	174.62	9.250	234.95	54	69		
Inboard/outboard marine	13.500	342.90	6.875	174.64	9.125	231.77	54	68		
Ordnance	11.187	284.14	10.562	268.27	9.031	229.38	53	69		
Tractors	11.875	301.62	6.875	174.62	10.750	273.05	62	78		
Lift trucks	18.000	457.20	7.125	180.97	9.625	244.47	71	91		
Heavy duty	21.000	533.40	7.500	190.50	10.000	254.00	89	115		
Tractors	16.000	406.40	7.125	180.97	11.000	279.40	75	102		
Heavy duty	20.500	520.70	8.687	220.34	10.000	254.00	112	130		
Diesel	20.875	530.22	8.750	222.25	9.531	242.08	91	115		
High duty	20.875	530.22	8.750	222.25	9.531	242.08	89	121		
Heavy duty	20.500	520.70	9.812	249.22	10.000	254.00	119	140		
High duty	20.875	530.22	11.000	279.40	9.531	242.08	114	153		
Heavy duty	20.500	520.70	11.000	279.40	10.000	254.00	129	160		
Heavy duty	20.500	520.70	11.000	279.40	10.000	254.00	129	160		
Ordnance	21.000	533.40	11.000	279.40	9.531	242.08	121	156		
High duty	20.875	530.22	11.000	279.40	9.531	242.08	114	148		
Diesel	20.875	530.22	11.000	279.40	9.531	242.08	117	153		

deliver 26 w/lb of battery. To produce 1,200 w will therefore require 1,200/26 = 46 lb of battery.

Unspillable lead-acid batteries of various types are not sealed but have most of the electrolyte retained in absorbent separators and have a vent trap which prevents leakage of electrolyte even in an inverted position. They are shown in Figs. 75 and 76.

Electrolyte-retaining batteries deliver high percentages of available power when operated inverted or in any position. Since the special insulators retain 80% of the electrolyte at the surface of plates, power producing capabilities remain at a high level. The percentage of capacity available from a typical electrolyte-retaining battery in various positions is 100% in the normal position, 90% at right angles, and 80% upside down.

Most types have built-in, ball-type specific gravity indicators. This simple device shows when the battery is fully charged and when a recharge is necessary.

Transparent plastic containers and covers make possible permanent leakproof seals and rugged, lightweight construction with minimum dimensions. Most of the 2-v units can be furnished as multiple cell batteries to provide higher voltages in various cell arrangements.

On the basis of suggested consumer prices, batteries for imported cars run about $1.00 to $1.20 per pound in the dry-charged state and about 80% of this on a weight basis when electrolyte is accounted for.

Aircraft batteries with spillproof vents run from $1.50 to $3.50 per

Fig. 75 Electrolyte-retaining battery. (*Wisco Division, ESB Incorporated.*)

Fig. 76 Electrolyte-retaining battery. (*Wisco Division, ESB Incorporated.*)

pound in the dry-charged state. Garden equipment batteries for tractors, golf carts, personnel carriers, and the like run about $1.10 to $1.50 a pound on the dry-charged basis, the higher values applying to spillproof vent construction. Batteries for 12-v inboard/outboard marine engines run about $.80 to about $1.10 per pound on the dry-charged basis, and $.60 to $.70 per pound on the complete electrolyte-filled wet basis.

The volume can be calculated from the weight, since automobile batteries will, in general, weigh 138 lb/cu ft, and the volume of the battery will be $46/138 = 0.333$ cu ft.

The rated capacity (the capacity at the 20-hr rate for a three-cell 6-v unit) can be estimated on the basis that a 100-ah battery will weigh 38 to 41 lb. Based on a 38-lb battery weight, a 46-lb battery will have a rated capacity of $46/38 = 120$ ah.

Dry-charged Batteries

Dry-charged lead-acid storage batteries are similar to wet lead-acid batteries except for additional processing steps, washing with water and drying in an inert or reducing atmosphere. Thus, the dry-charged battery is one containing formed positive plates of lead dioxide and formed negative plates of sponge lead together with dry separators. Dry-charged batteries are activated, or prepared for service, by adding battery-grade sulfuric acid as recommended by the manufacturer.

The advantages revolve around storage factors. Wet batteries deteriorate in storage. If recharged at frequent intervals, this deterioration is reduced. However, if wet batteries are not recharged frequently, they will wear out just as fast as if they were installed in automobiles and used every day. This useful service life of a dry-charged battery begins with the addition of the sulfuric acid.

The major standards are (1) 20-hr rate at 80°F, (2) 150- and 300-amp rates at 0°F, (3) charge rate acceptance at 30°F, (4) cycling life, and (5) overcharge life.

The 20-hr rate capacity is a measure of the long-term discharge capability and is closely related to the amount of active material within the battery. The battery is discharged at a constant, low-current value sufficient to run the battery completely down in 20 hr. The discharge is terminated when the average cell voltage drops to 1.75 v. The capacity is expressed in ampere-hours, being the product of the discharge current and the time duration of the discharge.

The cranking function is the battery's greatest responsibility. The improved generating capability of vehicle electrical systems through the use of self-rectifying ac generators has lessened the importance of the 20-hr rate.

A battery is also rated by specifying the voltage that it must deliver when discharged at a high current and low temperature. In the SAE standard, two discharge rates are used—150 and 300 amp and a temperature of 0°F. In Fig. 77, the difference in battery terminal voltage is shown when discharges are made at these two current values.

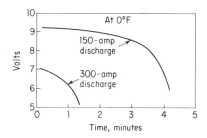

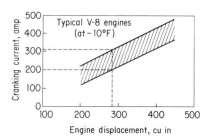

Fig. 77 Terminal voltage characteristics during discharge.

Fig. 78 Cranking current vs. engine displacement.

The high-rate discharge is meant to approximate the current drawn from the battery when the engine is cranked. The magnitude of the current is dependent upon such factors as cranking motor resistance and cranking ratio, engine bore and stroke, engine oil viscosity, and temperature. Figure 78 compares cranking current with engine displacement, keeping oil viscosity (SAE 10-W) and temperature ($-10°F$) constant. This relationship shows that the larger the engine, the greater the current drawn by the cranking motor for starting. The band reflects variation in design of the cranking motor, cranking ratio, etc. For instance, the cranking current can vary from 200 to 310 amp for engines of 280 cu in. displacement. The currents span from 100 to over 400 amp over the displacement range, and no one or even two current values used as ratings can relate the battery to its specific application.

The charge-rate acceptance test measures the current that a new, previously untested wet battery will accept at one-half charge at 30°F. A constant potential equal to 2.4 v per cell is used to simulate conditions in a vehicle with a voltage-regulated system. To meet the SAE standard, the battery must accept two times the 20-hr rate of discharge.

The life test is a series of charges and discharges to determine the ability of the battery to withstand a cycling application. The mode of failure is loss (shedding) of active material from the positive plates. The more shallow the discharge, the greater the number of cycles completed before battery failure (Fig. 79).

The overcharge life test conducted at constant current rather than constant potential (the case in an automotive vehicle) charges the battery at a constant current until the grids become so oxidized that the battery fails to maintain a 150-amp discharge for 30 sec. The failure is oxidation of the positive grids, carried out to the point where most of the grid material is converted to active material without strength, causing collapse of the plates.

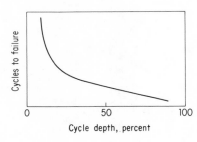

Fig. 79 Effect of cycle depth on cycling life.

Lowering the temperature increases the electrolyte viscosity, slowing the circulation between the plates. Figure 80 illustrates the reduction in the 20-hr rate capacity brought about by cooling the battery. The effect of low temperature is more severe for higher rates of discharge. Voltage-time curves on a typical automotive battery at 300-amp current drawn at 20°, 0°, and −20°F, are plotted in Fig. 81.

Increasing the discharge rate decreases the capacity that the battery will deliver owing to (1) loss of voltage (internal resistance), (2) diffusion of the electrolyte, limited by time, and (3) sulfation of the plate surfaces.

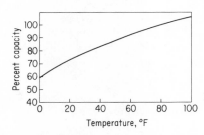 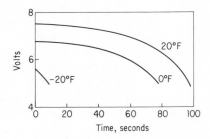

Fig. 80 Effect of temperature on low-rate discharge capacity (capacity at 80°F = 100%).

Fig. 81 Effect of temperature on high-rate discharge capacity (discharge rate 300 amp).

Figure 82 shows that ampere-hour capacity is a function of the discharge time or discharge rate. At the high rates of discharge, only a fraction of capacity can be obtained on the first discharge.

Cameron, Pettit, and Rowls[1] of Delco-Remy Division of General

Motors discussed batteries and cold cranking. The temperature effect is 100% available at 80°F, 62% at 32°F, and 40% at 0°F.

The effect of plate area on capacity is as shown in Fig. 83, and the effect of system losses in Fig. 84.

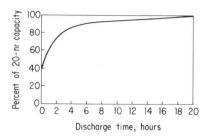

Fig. 82 Discharge capacity as a function of discharge time.

Hamm[2] of Delco-Remy Division of General Motors analyzed the cold starting problem of automobiles from the battery viewpoint and concluded that data from the American Automobile Association (AAA) indicate a cold starting problem is not confined to the extreme northern areas. Near the freezing point it is experienced to some extent by motorists in almost all states. Replacement batteries should show performance equal to the original equipment and the engine oil viscosity recommendations of the vehicle manufacturer should be followed.

Féry[3] has suggested a "dry accumulator" $Sn|dilute\ H_2SO_4|PbO_2$, filled with ceramic material and sealed, which gives a voltage of 1.75 v.

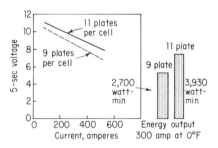 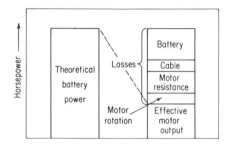

Fig. 83 Effect of plate area on battery capacity (60-ah battery at 0°F).

Fig. 84 Effect of system losses on power output.

Haring and Compton[4] studied the generation of SbH_3 by lead cells. Haring and Thomas[5] and Schumacher and Phipps[6] investigated the electrochemical behavior of lead, lead-antimony, and lead-calcium materials. They recommended the lead-calcium alloy containing 0.04 to 0.10% Ca

146   Batteries and Energy Systems

for storage-battery grids, in place of the widely used lead-antimony alloy, and for Planté plates, instead of lead. The lead-calcium cells retain their charges for longer periods than lead-antimony cells and are more efficient. The lead-calcium alloy shows a higher conductivity than the lead-antimony alloys. They have found application for "float" service in telephone exchanges. They are free from stibine (antimony hydride) evolution. Stibine is a poisonous hazard for "float service."

United Delco proposed an open-circuit voltage to test batteries. They suggest the application of an initial discharge to a battery to remove any surface charge which many defective cells are able to take and retain for a short period of time. Good discharge batteries will show considerable recovery immediately after the discharge load is removed. Defective batteries will always show less recovery. Since the voltage of a defective cell is lower than that of a normal cell, when the battery is placed on charge, either a much higher input current is absorbed and the final open-circuit voltage is much higher than the voltage of a good battery, or, if the defect is of such a type that the defective cell will not accept a charge, the final open-circuit voltage will be lower than a good battery. Delco has a cranking power rating in watts.

Counter Cells

Counter electromotive force (cemf) cells are similar to battery cells but are constant-voltage resistors. They consist of two groups of plates, but the plates are all alike and are "grids" without any active material. Neither group is positive or negative, although when direct current is passed through them they take on polarity just as a resistor does.

They maintain a constant voltage drop of approximately 2 v over a wide range of currents, whereas the voltage drop in a resistor is directly proportional to the current. They lower the voltage of a dc-source and maintain a constant voltage across a battery-supplied load when the battery voltage is raised for charging. As an example (see Fig. 85), assume a telephone installation with a system voltage of 48 v, and a permissible range of 46 to 50 v. Twenty-four cells supply voltage during an emergency discharge, yet the floating voltage for 24 cells is 51.5 v and for recharging at least 56 v. Therefore, three counter cells would be used, one in the circuit during normal floating operation and all during a recharge. A voltage-sensitive device accomplishes this automatically, and cuts the counter cells in or out successively as required to maintain a nearly constant load voltage.

As shown in Fig. 85, counter cells are connected in the load circuit, not the charge circuit. By short-circuiting them, there is no break or other interference in the load circuit.

The earlier counter cells were groups of lead-antimony plate grids, without any active material, and sulfuric acid electrolyte. The plate which acts as the positive scales within a comparatively short time, however, and also builds up a surface capacity which may cause arcing at the switch.

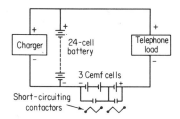

Fig. 85 Typical counter-cell application. (*ESB Incorporated.*)

In recent years, all such counter cells are made with either nickel or stainless steel plates and with an electrolyte of sodium hydroxide (NaOH). This type of cell does not build up any significant capacity, and the plates have an almost indefinite life. A $\frac{1}{2}$-in. layer of oil is floated on top of the electrolyte to reduce the spray from gassing and to minimize contact between the electrolyte and the air. The electrolyte requires renewal at long intervals.

Lead-Calcium Cells

A battery of flat or pasted-plate construction using grids of lead-calcium alloy has been developed. The percentage of calcium is very small, only enough to give the grids physical rigidity, thus it is nearly a "pure" lead grid. This alloy greatly reduces the local action or internal losses of the cell, but at the same time limits its use or application.

Such batteries in standby floating service draw less current owing to local action and thus require less frequent water addition. They will have a long life in years.

On the other hand, the nearly pure lead grid of the positive plate is more susceptible to "formation" (corrosion) from charging. All unnecessary charging is avoided. Any regular cycle type of operation as the regular rechargers would soon "form" the grid to the point where it would have high resistance and crack and crumble.

They also develop a higher voltage near the end of charge, which means that in order to charge them fully, either a higher charger voltage or a longer time is required.

These batteries are limited to standby service, either on float or open circuit where the necessary charging is minimal.

An average storage battery contains better than 23 lb lead per unit and in service shows a life greater than 2 years. Harned and Hamer[7] reported a summary of the molal electrode potentials and the reversible emfs of the lead accumulator from 0 to 60°C. Storage-battery life tests were discussed by Hatfield and Harner.[8]

W. A. Cunningham[9] described new designs for charge-retaining batteries. "Charge-retaining" or "CR"-type lead-acid secondary cells and batteries have been used since 1933 by the Lighthouse Service and U.S. Coast Guard for lighted buoys and minor lights on fixed objects.[10] Charge-retaining cells and batteries retain 85 to 90% of their rated capacity during a 1-year open-circuit stand at 80°F. Automotive lead-acid batteries exhibit a 10 to 15% loss of rated capacity in approximately 3 weeks. Charge-retaining batteries are shown in Figs. 86, 87, and 88.

The charge retention is achieved by keeping all deleterious impurities to the lowest practicable minimum, particularly antimony and iron in all internal components. Grids for lead-acid batteries are cast from an antimony-lead alloy. Antimony accelerates self-discharge of the battery by reacting with the sponge-lead active material of the negative plates. If

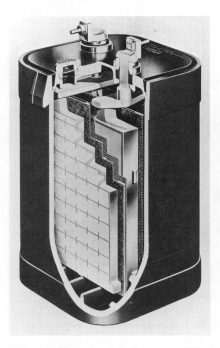

Fig. 86 Cutaway of charge-retaining battery. (*Wisco Division, ESB Incorporated.*)

Fig. 87 Single cell, charge-retaining battery. (*Wisco Division, ESB Incorporated.*)

Fig. 88 External view of charge-retaining battery. (*Wisco Division, ESB Incorporated.*)

pure lead grids are used, the self-discharge reaction is reduced substantially. By using pure lead for grids in the system, self-discharge is reduced to a low level.[11,12]

In 1955, development of a 3,000-ah, 6-v CR battery for anticipated use in buoy service was initiated. This battery shown in Fig. 89, consisting

Fig. 89 Battery with massive plates, plastic grids, and contact with the active material by means of $\frac{1}{4}$- and $\frac{1}{8}$-in. lead wires. (*Wisco Division, ESB Incorporated.*)

of three series-connected cells enclosed in a cylindrical steel container, was intended to replace a series-parallel combination of fifteen 600-ah, 2-v CR cells. Based on the weight of the grid in the 600-ah cell, a single pure-lead grid for the CR-6-3,000 battery would weigh approximately 18 lb. Experimental design, construction, and processing studies were undertaken with plastic grid frames incorporating pure lead wires to provide electrical conductivity. The completed polystyrene grid weighed only 5.8 lb, an approximate saving of 12.2 lb per grid.

As each battery required the equivalent of 18 full-thickness grids (five full-thickness, two half-thickness grids per cell), a weight reduction of 220 lb, or 14% of battery weight, was realized.

Full-thickness grids were assembled from sections of polystyrene approximately 34 in. long, $8\frac{3}{4}$ in. wide, and $\frac{1}{2}$ in. thick. Two of these half-sections were cemented together, with four $\frac{1}{4}$-in. diameter pure lead conductor wires traversing the interior length of the grid. The conductor wires were bent into a sinusoidal configuration in order to traverse the full thickness of the grid in each compartment. The upper ends of the wires were "burned" to an antimony-lead alloy terminal (low antimony content), and the bottom ends were secured in slots molded into the grid.

TABLE 37 CR-6-3,000 Battery Data

Approximate overall dimensions	21 in. diameter, 59 in. high
Weight range (including watertight steel cover)	1,350 to 1,385 lb
Weight of steel container and cover assembly	360 to 365 lb

	Ampere-hour capacity to:			
Type of discharge	5.85 v	5.70 v	5.40 v	5.10 v
(A) Continuous at 80°F:				
0.406 amp	3,137	3,546	4,043	
0.812 amp	2,952	3,391	3,722	
1.624 amp	2,709	3,274	3,644	
5.0 amp	2,787	3,274	3,668	
(B) Intermittent				
3 min at 50 amp, 2 hr, 57 min rest, repeat cycle until voltage on discharge reaches 5.10 v. Temperature 80°F	1,175	2,000	2,675	3,225
(C) 5 amp continuous discharge at 80°F before and after 2 years open-circuit stand at 80°F:				
Full-charge capacity (at 5 amp discharge)	2,787	3,274	3,668	
Capacity after 2 years stand at 80°F	1,825	2,400	2,785	
Loss of capacity, 2 years	962	874	883	
Indicated loss of charge per year	17.3%	13.4%	12.1%	

Half-thickness grids, for "end" or "outside" negatives, were assembled in a similar manner, utilizing only one polystyrene half-thickness section. The performance is given in Table 37.

In 1966, Wisco Division of ESB was awarded a contract by the U.S. Navy Bureau of Weapons for a 1,000-ah CR battery as a power supply for Project NOMAD. This was accomplished by designing the plates, cell containers, and outer steel containers to roughly one-third the height of those used in the CR-6-3,000 battery, and in June, 1967, sixteen 6-v CR-6-1,000 batteries were delivered to the Navy. These 6-v units will be nested in pairs and series-connected to provide eight 12-v batteries for a NOMAD barge.

Passenger automobiles have grown from about 60 million in 1959 to over 80 million in 1967, and the total number of cars, buses, and trucks from about 71 million in 1959 to almost 97 million in 1967. Scrappage is at the rate of about 7 million vehicles per year. Batteries, replacement, original equipment, and export totaled about 32 million replacements,

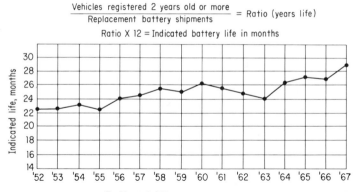

Year	Ratio	Indicated life, months	Year	Ratio	Indicated life, months
1942	1.86	22.3	1955	1.87	22.4
1943	1.78	21.4	1956	2.01	24.1
1944	1.58	19.0	1957	2.06	24.7
1945	1.73	20.8	1958	2.14	25.7
1946	1.80	21.6	1959	2.08	25.0
1947	1.21	14.5	1960	2.18	26.2
1948	1.30	15.6	1961	2.10	25.2
1949	1.77	21.2	1962	2.04	24.5
1950	1.47	17.6	1963	2.01	24.1
1951	1.73	20.8	1964	2.22	26.6
1952	1.88	22.6	1965	2.29	27.5
1953	1.89	22.7	1966	2.26	27.0
1954	1.92	23.0	1967	2.39	28.6

Fig. 90 Indicated battery life. (*The Association of American Battery Manufacturers, Inc.*)

152   Batteries and Energy Systems

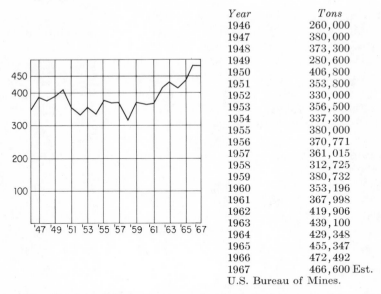

Year	Tons
1946	260,000
1947	380,000
1948	373,300
1949	280,600
1950	406,800
1951	353,800
1952	330,000
1953	356,500
1954	337,300
1955	380,000
1956	370,771
1957	361,015
1958	312,725
1959	380,732
1960	353,196
1961	367,998
1962	419,906
1963	439,100
1964	429,348
1965	455,347
1966	472,492
1967	466,600 Est.

U.S. Bureau of Mines.

Fig. 91 Total lead used by battery manufacturers, in thousands of tons (includes primary, secondary, and antimonial lead in tons of 2,000 lb). (*The Association of American Battery Manufacturers, Inc.*)

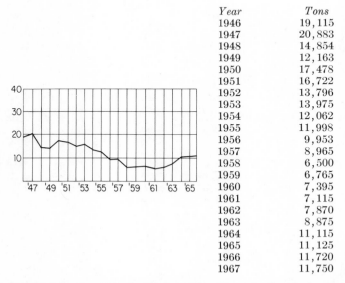

Year	Tons
1946	19,115
1947	20,883
1948	14,854
1949	12,163
1950	17,478
1951	16,722
1952	13,796
1953	13,975
1954	12,062
1955	11,998
1956	9,953
1957	8,965
1958	6,500
1959	6,765
1960	7,395
1961	7,115
1962	7,870
1963	8,875
1964	11,115
1965	11,125
1966	11,720
1967	11,750

Fig. 92 Red lead purchased by battery manufacturers, in thousands of tons. Figures previous to 1957 from U.S. Bureau of Mines, thereafter from Lead Industries Association. (*The Association of American Battery Manufacturers, Inc.*)

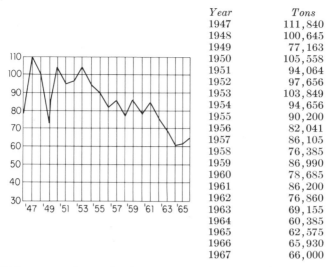

Year	Tons
1947	111,840
1948	100,645
1949	77,163
1950	105,558
1951	94,064
1952	97,656
1953	103,849
1954	94,656
1955	90,200
1956	82,041
1957	86,105
1958	76,385
1959	86,990
1960	78,685
1961	86,200
1962	76,860
1963	69,155
1964	60,385
1965	62,575
1966	65,930
1967	66,000

Fig. 93 Litharge purchased by battery manufacturers, in thousands of tons.* Figures previous to 1957 from U.S. Bureau of Mines, thereafter from Lead Industries Association. (*The Association of American Battery Manufacturers, Inc.*)

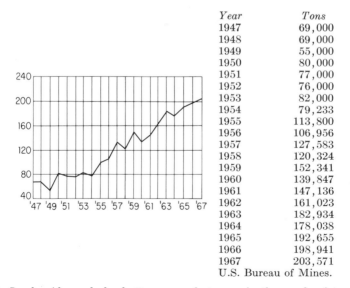

Year	Tons
1947	69,000
1948	69,000
1949	55,000
1950	80,000
1951	77,000
1952	76,000
1953	82,000
1954	79,233
1955	113,800
1956	106,956
1957	127,583
1958	120,324
1959	152,341
1960	139,847
1961	147,136
1962	161,023
1963	182,934
1964	178,038
1965	192,655
1966	198,941
1967	203,571

U.S. Bureau of Mines.

Fig. 94 Lead oxide made by battery manufacturers, in thousands of tons. (*The Association of American Battery Manufacturers, Inc.*)

* Previous to 1957 from U.S. Bureau of Mines. Thereafter from Lead Industries Association.

9 million original equipment, and a quarter of a million exports (these exports had a value of better than 4 million dollars), a total of nearly 42 million units. Imports of motor vehicles in 1967 were better than 1 million units.

Indicated battery life is shown in Fig. 90. Total lead used by battery manufacturers is shown in Fig. 91, red lead purchased in Fig. 92, litharge in Fig. 93, and oxide made by battery makers in Fig. 94.

REFERENCES

1. G. L. Cameron, C. W. Pettit, and G. A. Rowls (Delco-Remy Div., General Motors Corp.), Cold Cranking Team Battery, Cables, Cranking Motor, Engine Oil, Paper No. 894-B, *SAE, National Farm, Construction and Industrial Machinery Meeting*, Milwaukee, Wisc. (Sept. 14–17, 1964).
2. A. A. Hamm (Delco-Remy Div., General Motors Corp.), Cold Facts on Cold Starting, Paper No. 885A, *SAE, Summer Meeting*, Chicago, Illinois (June 8–12, 1964).
3. C. J. V. Féry, French patent 748,630.
4. H. E. Haring and K. G. Compton, *Trans. Electrochem. Soc.*, **68**: 283 (1935).
5. H. E. Haring and U. B. Thomas, *Trans. Electrochem. Soc.*, **68**: 293 (1935).
6. E. E. Schumacher and G. S. Phipps, *Trans. Electrochem. Soc.*, **68**: 309 (1935).
7. H. S. Harned and W. J. Hamer, *J. Am. Chem. Soc.*, **57**: 33–35 (1935).
8. J. E. Hatfield and H. R. Harner, *Trans. Electrochem. Soc.*, **71**: 583, 597 (1937).
9. W. A. Cunningham, New Design Charge-retaining Batteries, *Marine Tech. Soc. J.*, Washington, D.C. (1967).
10. C. C. Rose and A. C. Zachlin, Low-discharge Cells, *Trans. Electrochem. Soc.*, **68**: 273 (1935).
11. C. C. Rose and A. C. Zachlin, Low-discharge Cells, *Trans. Electrochem. Soc.*, **99**: 9, 243C (1952).
12. C. F. Gerhan, E. A. Kure, and H. B. Grohe, Charge-retaining Cells, *Sixth International Lighthouse Conference*, Washington, D.C. (Sept. 1960).

chapter 14

Alkaline Secondary Cells

Although the lead-acid battery had distinct advantages over dry, or primary, batteries and could be used in wider applications than the dry battery, many features of the ideal storage battery were missing from the lead-acid system.

About the turn of the century, Edison in the United States and Jungner in Sweden were working on storage batteries with alkaline electrolytes, Edison with the nickel-iron couple and Jungner with the nickel-cadmium couple. Whereas Edison had envisioned a nickel-iron battery in every electric automobile, Jungner developed nickel-cadmium batteries for hand lanterns, fire-alarm systems, and railroad applications.

Type and Theory

Alkaline storage cells in the United States, other than nickel-cadmium (treated separately), are the Hubbell, of the system Ni threads and Ni| oxide|KOH|Fe, for miners' lamps, and the Edison, of the system of finely divided Ni + Ni peroxide|21% KOH|finely divided Fe. The active materials of the Edison cell are nickel peroxide for the positive plate and finely divided iron for the negative. Cell construction is shown in Fig. 95. Small amounts of LiOH are added to the electrolyte, and mercury is incorporated with the iron of the negative plate to overcome the iron passivity. The reactions are

$$\text{Discharge} \rightarrow$$
$$8KOH + 6NiO_2 + 3Fe \rightleftharpoons Fe_3O_4 + 2Ni_3O_4 + 8KOH$$
or $\quad 6NiO_2 + 3Fe \rightleftharpoons Fe_3O_4 + 2Ni_3O_4$
$$\leftarrow \text{Charge}$$

The major manufacturer states, "In a fully charged cell the active materials are essentially nickel oxyhydrate (NiOOH) and metallic iron sponge (Fe) and the electrolyte potassium hydroxide (KOH) and water

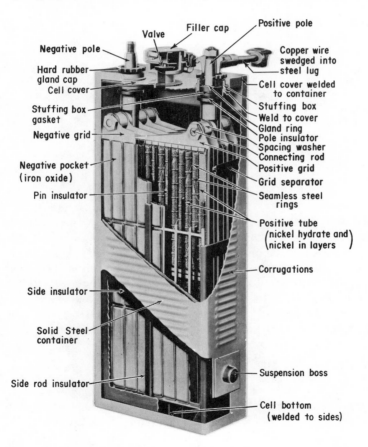

Fig. 95 General construction of the Edison cell. (*ESB Incorporated.*)

with the addition of lithium hydroxide. The chemical reactions during discharge produce intermediate transitory compounds, but the end result is a transfer of oxygen, leaving the positive as nickel hydroxide $Ni(OH)_2$ and the negative as iron hydroxide $Fe(OH)_2$. Neglecting the intermediate reactions, the end formula may be considered as

$$2NiOOH \cdot H_2O + Fe \underset{\leftarrow \text{Charge}}{\overset{\text{Discharge} \rightarrow}{\rightleftharpoons}} 2Ni(OH)_2 + Fe(OH)_2$$

The electrolyte takes no part in the chemical reaction in the sense that its composition is changed." It acts as a transfer agent by "accepting" oxygen at the positive during discharge and "depositing" it at the negative. On charge, conversely, the action is reversed and the positive and negative materials revert to their original charged state.

When the charging current is greater than the cell can "absorb," the excess is expended in decomposing the water into hydrogen and oxygen, given off as gases. This is the reason for the addition of water.

The active material in both cases is in the form of finely ground powder and requires a container to give it a physical form and provide a conductor to the external circuit.

The active materials are encased in steel tubes or pockets. These are held in position in a sheet steel frame to form a plate. These frames have an extension at the top in the form of a lug or ear in which there is a hole for a connecting rod. Plate lugs with intermediate spacing washers are placed over the rod and retained by nuts at either end. On the rod, at the center, is the lower part of the post which extends upward as the external connection to the circuit. This assembly of plates, rod, and post is a group.

The plates of the positive and negative groups are intermeshed and prevented from coming in contact by round hard rubber or plastic pins or insulators. This assembly of positive and negative groups and insulators is an element.

The container is nickel-plated sheet steel insulated internally from the element and from the posts where they pass through the top or cover. Hard rubber or plastic sheet and hard rubber bushings or gland caps encircle the posts.

Each cell must be insulated from the adjoining ones. The sides of each container may be equipped with small round extension bosses which project into rubber buttons in holes in the sides of a wood crate or tray. The cell is suspended from these bosses, thus insulating it from adjoining cells. The bottom of the tray is open for ventilation.

The tray and the sides and bottom of the cells are covered with insulating paint, and the top of the cells with a resinous coating.

The voltage of nickel-iron cells is 1.2 v, thus a 10-cell battery is a 12-v battery, a 30-cell battery a 36-v battery, etc. The voltage, however, will depend on whether it is on open-circuit, on discharge, or on charge, and in the latter cases, the current rate and the state of charge.

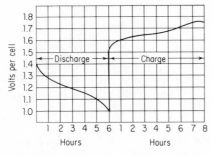 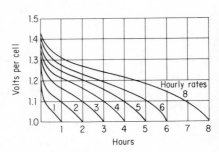

Fig. 96 Typical voltage characteristics during a constant-rate discharge and recharge. (*ESB Incorporated.*)

Fig. 97 Time-voltage discharge curves, at various rates, to a "final" voltage of 1.0 v per cell. (*ESB Incorporated.*)

The open-circuit voltage will vary between 1.25 and 1.35 v, but its relation to the state of charge is not sufficiently definite to be an indication of condition.

When on discharge, the voltage will decrease to a lower value owing to the internal resistance. This "drop" will increase with an increase in discharge current. At any constant rate of discharge, the voltage decreases until as the cell approaches exhaustion, it falls below a value where it is no longer effective, varying with the rate of discharge. The value is 1.0 v per cell. See Figs. 96 and 97 for discharge curves at various rates.

Conversely, when placed on charge, the voltage rises sharply to a maximum value varying with the charging rate. At most rates, this maximum is 1.7 to 1.8 v per cell. See Fig. 98 for typical charge-voltage curves.

The electrolyte is a potassium hydroxide (KOH) solution with the addition of lithium hydroxide. It is not significantly changed throughout the normal range of charge and discharge.

The specific gravity of the electrolyte of a new cell is about 1.210 to 1.215 at normal electrolyte level, at 77°F. When the gravity decreases to about 1.160, it affects the capacity and the operation of the cell, and the solution requires renewal. This may be necessary several times during the life of the battery.

Significant change in the specific gravity occurs if the discharge be carried down to or near zero voltage, as when cells are in storage. The lithium ion moves from the plates into the electrolyte, raising the specific gravity 25 to 35 points (.025 to .035), reversing when the cells are charged.

The higher the discharge rate, the fewer ampere-hours a battery will deliver under similar conditions, as in Fig. 99. The primary reason is

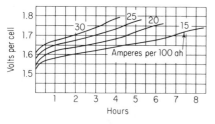

Fig. 98 Typical charging voltages at various current rates. (*ESB Incorporated.*)

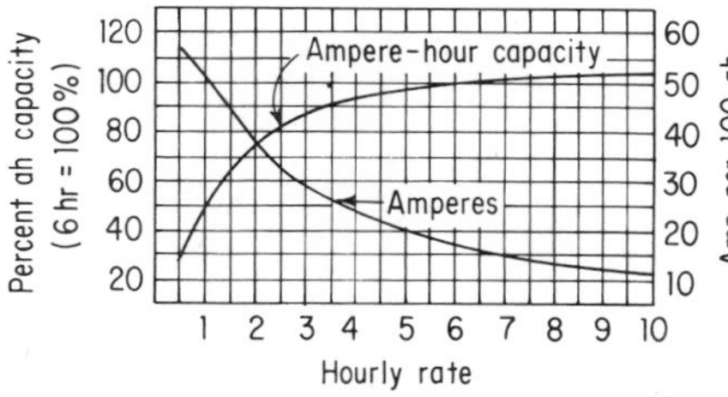

Fig. 99 Capacity-rate curves to a final voltage of 1.0 v per cell at 77°F. (*ESB Incorporated.*)

the internal resistance of the cell. The higher the current, the greater is this drop and, therefore, the less voltage is available for the external load. The final voltage is reached earlier with a loss in capacity.

The resistance and viscosity of the electrolyte are reduced at higher temperatures. Thus the drop within the cell is less, capacity is greater. Conversely, at lower temperatures the capacity is reduced. Figure 100 gives a general indication of this temperature effect at various discharge rates.

Figures 97 and 99 show that the ampere-hour capacity to a final voltage decreases with an increase in current rates. The voltage falls to its final value in a shorter period of time. If, after a high-rate discharge to a given

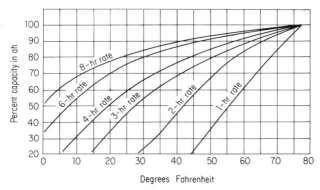

Fig. 100 Effect of temperature on capacity at various rates, 77°F = 100%. (*ESB Incorporated.*)

voltage, the current be reduced, the cell voltage will recover and further capacity may be obtained at the lower rate before the voltage again drops to the same value. When decreasing current rates are used, the total ampere-hour capacity of a battery at a given hourly rate may be obtained

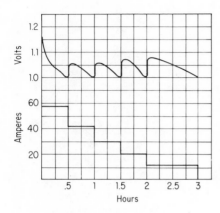

Fig. 101 Effect of decreasing discharge rates on cell voltage. (*ESB Incorporated.*)

during that time period, regardless of higher-than-average rates at the beginning. This does not hold (Fig. 101) if the high rates are at the end of the discharge.

During discharge there is a rise in temperature owing to the heat losses (I^2R) in the cell, this increasing with an increase in current rates. During discharge of 6 to 8 hr, this may cause a rise of 20 to 30°F or more if reasonable ventilation be lacking. The discharge should be limited to 80% of capacity.

Edison cells may be employed for ignition and lighting of gasoline motor cars, but because of their high internal resistance are not used for motor starting. The fulfillment of all three needs for the motor car is well met by the lead storage battery. Alkaline cells find application in electric vehicles, storage-battery street cars, mining locomotives, industrial trucks, miner's lamps, and underground sources of power. They are not used for load regulation in power systems because of their heavy voltage drop at high discharge rates. The commercial cells show ampere-hour efficiencies of 82% and watt-hour efficiencies of 60%, with an average capacity of about 13 wh/lb of cell.

Alkaline–Manganese Dioxide

Alkaline–manganese dioxide secondary batteries are maintenance free, hermetically sealed, and will operate in any position. They have been

TABLE 38   Characteristics of Alkaline Secondary Batteries

Cell size	Nominal voltage	Average operating voltage	Rated ampere-hour capacity	Max. recommended discharge current, amp	Recommended charging current (constant current), amp
D	1.5	1.0–1.2	2.0	0.5	0.25
F	1.5	1.0–1.2	3.2	0.8	0.4
G	1.5	1.0–1.2	4.0	1.0	0.5
Eveready:					
563	4.5	3 D	2	0.5	0.25
560	7.5	5 D	2	0.5	0.25
564	13.5	9 G	4	1	0.5
561	15.0	10 G	4	1	0.5

designed for electronic and electrical applications where low initial cost and low operating cost are paramount.

The total number of times the alkaline–manganese dioxide secondary battery can be recharged is fewer than that of the nickel-cadmium system, but the initial cost is lower.

The individual cells use electrodes of zinc and manganese dioxide with an electrolyte of potassium hydroxide. These are put together in an inside-out cell construction and each cell is hermetically sealed. Each cell has a nominal voltage of 1.5 v. They are made in three sizes—D, F, and G. Basic characteristics of these cells are shown in Table 38. A cutaway view of the cell is shown in Fig. 102.

Finished batteries are constructed by connecting the required number of the proper cell size in series and sealing them in a metal case. Present

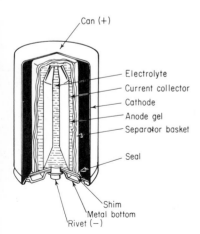

Fig. 102   Cutaway of alkaline secondary cell. (*Eveready, Union Carbide Corporation.*)

types include 4.5- and 7.5-v batteries using size D cells and 13.5- and 15-v batteries made up of G cells. Specifications are listed in Table 20, giving dimensions and related data on the alkaline zinc–manganese dioxide units. Charging data is given in Table 38.

The discharge characteristic of the secondary cell is similar to that of the primary. The voltage decreases slowly as power is withdrawn from the battery. The shape of this discharge curve changes slightly as the battery is repeatedly charged and discharged. The total voltage drop for a given power withdrawal increases as the number of charge and discharge cycles increases (see Fig. 103).

When cells are discharged for 4 hr at the maximum discharge current and then recharged at constant current for 10 hr, the cycle can be repeated 25 to 35 times before falling below 0.9 v per cell in any 4-hr discharge period. Decreasing either the discharge current or the total ampere-hour withdrawal will increase the cycle life. If the power demands be increased, the cycle life will decrease.

During the early part of its cycle life there is a power reserve amounting to 100 and 200% of the rated capacity. Terminal voltage may be 1.0 to 1.2 v per cell after delivery of rated ampere-hour capacity. If discharged beyond rated capacity, cycle life will be reduced. Figure 103 shows the minimum voltage that a battery will reach at the end of the discharge period and the maximum voltage during charge period.

A new unit is shipped charged and has the retention characteristics of primary batteries. It must be discharged to its rated capacity before it will be capable of standing any overcharge. The unit must be discharged before recharging it. Recharging applies the current for a time long enough to replace about 125% of the ampere-hours removed.

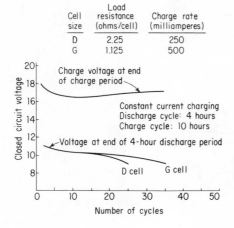

Fig. 103 Discharge and charge voltage characteristics of an alkaline–manganese dioxide secondary battery containing ten D- or G-size cells. (*Eveready, Union Carbide Corporation.*)

chapter 15

The Nickel-Cadmium System

The active material of the positive plate is nickel hydrate, with graphite aiding conductivity. The active material of the negative plate is cadmium sponge, with additives to aid conductivity.

The electrolyte is a solution of potassium hydroxide (KOH), the gravity of which is normally 1.160 to 1.190 (depending on type of cell) at 77°F. A small amount of lithium hydroxide is included to improve capacity.

The charge-discharge reaction may be written

$$2Ni(OH)_3 + Cd \rightleftharpoons 2Ni(OH)_2 + Cd(OH)_2$$
$$\text{charged} \leftrightarrow \text{discharged}$$

and is shown in Fig. 104.

Most of the engineering developments of the Edison cell have been applied to the Exide form of the nickel-cadmium unit. When fully charged, the nickel hydrate of the positive plate is oxidized, while the active material of the negative plate is metallic cadmium sponge. When the battery is discharged, the positive-plate active material is reduced to a lower oxide, while the metallic cadmium sponge of the negative plate is oxidized.

164   *Batteries and Energy Systems*

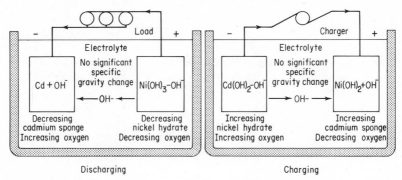

Fig. 104 Chemical action in a nickel-cadmium storage battery. (*ESB Incorporated.*)

The reaction consists of a transfer of oxygen ions from one set of plates to the other, the electrolyte acting as transfer agent. As the electrolyte takes no part in the chemical reaction, its specific gravity does not change materially during charge or discharge.

The positive and negative plates of the nickel-cadmium battery are pocket type, their basic construction being identical. The pockets are formed from very thin strips of nickel-plated steel, finely perforated (Fig. 105).

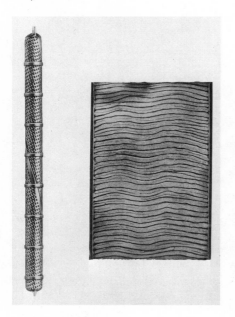

Fig. 105 Perforated steel-strip pockets containing active material. (*ESB Incorporated.*)

To make a pocket, a strip is formed into a trough by passing it through a pair of rollers. At the same time, active material is fed into the trough. A similar perforated strip is superimposed, and the edges of the two strips are crimped together. This forms a long, narrow pocket, with the powdered active material held firmly inside.

Individual pockets, as many as necessary to produce the required plate height, are attached to one another by locking the crimps. They are then cut to the required plate width. This assembly is a "plaque."

A U frame is placed around the edges of the plaque and pressed into place by rolls. The rolls also press insulator-pin grooves into the face of the now completed plate.

Plates of the same polarity are attached to their respective terminal poles by a steel bolt and locknut. Steel washers assure spacing of the plates. This assembly is a group. (In small cells, plates are welded to the terminal poles.)

Positive and negative groups are then intermeshed and pin insulators put in place. This assembly is an element (Fig. 106) and is inserted into the cell container of steel or plastic. A cell cover with vent cap and holes for terminal completes the assembly (Fig. 107).

The electrolyte reservoir above the plates should be large enough to

Fig. 106 "Element" of battery. (*ESB Incorporated.*)

Fig. 107 Cell cover with vent. (*ESB Incorporated.*)

reduce the frequency of adding water. This has become mandatory under the electrical codes of some states.

The nominal voltage of a nickel-cadmium cell on discharge is 1.2 v. The voltage of a nickel-cadmium cell depends on whether the cell is on

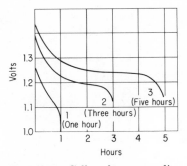

Fig. 108 Cell voltage on discharge at 1-, 3-, and 5-hr rates. (*ESB Incorporated.*)

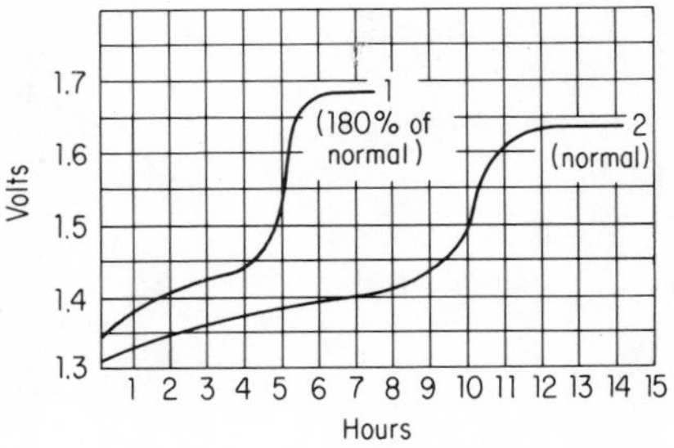

Fig. 109 Cell voltage on charge at 180% of normal rate and at normal rate. (*ESB Incorporated.*)

open circuit, charge, or discharge. Open-circuit voltage may vary from 1.30 to 1.38 v and is not an accurate indicator of the state of charge.

When the cell is connected to an external load, its voltage will fall to a value dependent on discharge rate and state of charge. If the rate be constant, the voltage decreases until it reaches a point where the cell is discharged. A final voltage is 1.14 v per cell.

Figure 108, showing three discharge curves, illustrates voltage drop at various rates. Curve 1 shows the voltage of a cell discharging at five times the normal rate to 1.0 final volts (5-hr rate considered "normal rate" in this example). Curve 2 shows the voltage of a cell discharging at 160% of normal rate to 1.10 final volts, and curve 3 shows voltage of a cell discharging at the 5-hr rate to 1.14 final volts.

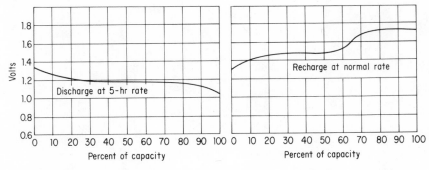

Fig. 110 Typical discharge-recharge characteristics. (*ESB Incorporated.*)

TABLE 39 Freezing Points and Gravity of Electrolyte

Temperature, °F	Specific gravity, corrected to 77°F
0	1.150
−10	1.180
−20	1.205
−30	1.225
−40	1.240

When a cell is placed on charge, its voltage rises rapidly at first, then slows to a more moderate rate of increase until the charge is approximately 70% complete. At this point, the voltage rises rapidly again, even more rapidly than at the beginning of the charge. Finally, it levels off and becomes practically constant.

Figure 109 shows how the voltage curves vary with the rate of charge. Curve 1 shows cell voltage while charging at 180% of normal rate. Curve 2 shows cell voltage while charging at the normal rate.

Figure 110 summarizes the voltage characteristics of a nickel-cadmium cell, showing a typical discharge-charge curve at normal rates.

The specific gravity of a new cell (depending on type) is 1.160 to 1.190 at 77°F and at the normal solution level. As no constituents of the electrolyte combine with the active material of the plates, the specific gravity does not change appreciably during charge or discharge, nor does it indicate the state of charge or discharge.

As electrolyte temperatures drop to sub-zero, ice crystals form. However, the electrolyte will not freeze unless its corrected specific gravity is below that in Table 39.

Figure 111 gives the general picture of capacity.

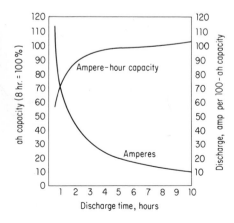

Fig. 111 Capacity vs. time curve to 1.14 per cell final voltage at 77°F. (*ESB Incorporated.*)

Figure 111 (and also Fig. 112) indicates that ampere-hour capacity to a given final voltage decreases with an increase in current rates. This is not a loss of capacity. Final voltage is reached in less time.

A nickel-cadmium battery may be charged at any rate that does not cause cell temperature to exceed 120°F.

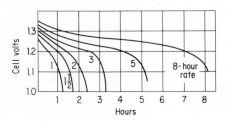

Fig. 112 Time-voltage discharge curves at various rates. (*ESB Incorporated.*)

Under normal conditions, the cells will start gassing when the voltage reaches 1.45 to 1.47 v, indicated by the "popping" of the valves in the vent caps of steel cells, and the rattling of the balls in plastic cells. Gassing increases water consumption. The charging rate is reduced to normal when gassing starts. The normal rate is usually 20 amp/100 ah at the 8-hr rate.

An ampere-hour input approximately 40% greater than the preceding discharge is necessary to return the battery to its previous state of charge. Insufficient charging results in lowered capacity, while excessive charging results in abnormal water loss.

Unsealed and Sealed Cells

The nickel-cadmium cell has been used in Europe for many years as an unsealed cell. The extension of the system has been to small hermetically sealed batteries free of the addition of water.

Sealed nickel-cadmium cells can be recharged many times and are not adversely affected by long standing, either charged or discharged. Portable devices require more energy than is economically available from primary batteries. These are electronic photoflash, dictating machines, electric shavers, tape recorders, instruments, alarm systems, transistor radios, transmitters, receivers, movie cameras, emergency lighting, hearing aids, amplifiers, telemetering, and tools and appliances.

A vented nickel-cadmium battery will liberate oxygen and hydrogen plus entrained electrolyte fumes through a valve. To hermetically seal a nickel-cadmium cell, a means of using up this gas inside the cell employs excess ampere-hour capacity in the cadmium electrode. With both electrodes in the discharged state, charging causes the nickel electrode to reach full charge first and oxygen is generated. The cadmium electrode has not yet reached full charge so it cannot cause hydrogen to be generated. The oxygen formed can reach the surface of the metallic cadmium electrode where it reacts, forming cadmium oxide. In overcharge,

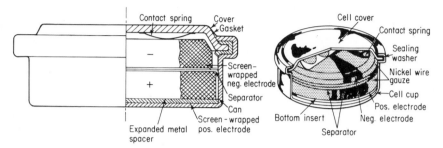

Fig. 113 Cutaway views of button cell. (*Eveready, Union Carbide Corporation.*)

the cadmium electrode is oxidized at a rate sufficient to offset input energy, keeping the cell in equilibrium at full charge.

Nickel-cadmium cells are available in button (20–3,000 mah), cylindrical (450–8,000 mah), and rectangular (1.6–23 ah) configurations and capacities.

Cutaway views of a standard-rate button cell are shown in Fig. 113. A cross section of a double-plate, molded-electrode, high-rate button cell is illustrated in Fig. 114.

Button-type cells utilize a cell cup (positive pole) and a cell cover (negative pole). The cells are available with and without solder tabs.

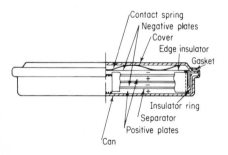

Fig. 114 Cross-section view—double plate, molded-electrode, high-rate button cell. (*Eveready, Union Carbide Corporation.*)

The electrodes consist of tablets wrapped in nickel wire gauze, separated by a fine-pored separator. Sealing is accomplished by flanging the rim of the cell cup over the rim of the cell cover with a plastic washer between, serving to insulate the cup from the cover.

170 Batteries and Energy Systems

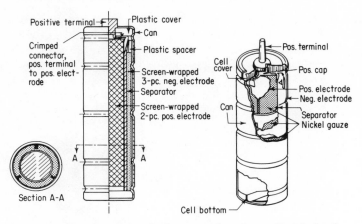

Fig. 115 Cutaway of standard-rate, pocket-plate cylindrical cells. (*Eveready, Union Carbide Corporation.*)

When required, two to ten button cells may be assembled into a higher-voltage series stack.

Internal construction features for cylindrical cells (Fig. 115) consist of a corrugated steel case housing the positive electrode, which is a split cylinder, as well as the negative electrode in the form of three shell-shaped segments laid around the positive electrode. The positive electrode as well as the individual negative segments are wrapped in nickel gauze. For insulation, a porous separator is placed between the electrodes. The

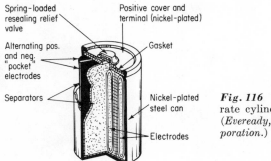

Fig. 116 Cutaway of standard rate cylindrical cell (type C2). (*Eveready, Union Carbide Corporation.*)

negative electrode rests against the metal case with which it holds firm contact.

Another construction (used in No. C3) is shown in Fig. 116. Electrodes are porous sintered plates to which the active materials are applied. A thin porous nickel plaque is made by firing fine nickel powder at high temperature until the particles sinter together forming a porous structure. This structure is sintered around a nickel screen. The plaque serves as a supporting structure. The plates are processed electrochemically so that active nickel and cadmium oxides are deposited within the pores. This provides low internal resistance.

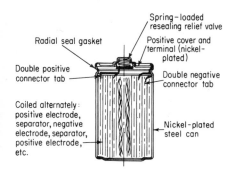

Fig. 117 Cutaway of typical cylindrical sintered plate cell. (*Eveready, Union Carbide Corporation.*)

A cutaway drawing of a cylindrical sintered plate cell is shown in Fig. 117.

The cell case of a rectangular cell is welded, plated steel. The case has a safety vent. Connections are made to a nut and stud negative terminal and a welded tab positive terminal in the smaller cell sizes, and to a nut and stud for both positive and negative terminals in the larger cell sizes. The steel case is polarized positive for all rectangular cells. A cross section of the rectangular, sintered plate cell is illustrated in Fig. 118. Electrodes consist of porous sintered nickel plaques into which the active materials are impregnated.

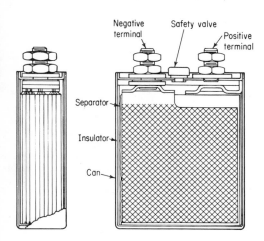

Fig. 118 Cutaway of typical high-rate, rectangular, sintered plate cell. (*Eveready, Union Carbide Corporation.*)

The best material for electrical contact to nickel-cadmium cells is nickel-plated steel. The nickel plating should be directly on the steel base, not over copper or other active materials. Plating is 0.0005 to 0.001 in.

The capacity rating of nickel-cadmium batteries is based upon a discharge period of 10 hr (10-hr rate) and an end-point voltage of 1.1 v per cell, or a discharge period of 1 hr (1-hr rate) and an 0.9-v end point.

The open-circuit voltage of a sealed nickel-cadmium cell at room temperature ranges from 1.4 to 1.28 v, depending on how long the cell has been standing after charge. During discharge, the average voltage is approximately 1.2 v per cell. At normal discharge rates the characteristic is nearly flat until complete discharge is approached. The battery provides most of its energy above 1.1 v per cell.

Self-discharge characteristics are shown in Fig. 119 as a decline in percent capacity. After the first month the capacity decreases gradually with age. High-discharge-rate cells lose charge on shelf.

Sealed cells show a small loss of capacity at temperatures from -20 to $+45°C$.

Nickel-cadmium batteries have long been proposed for automotive use, but their higher cost has not been helpful in making them competitive to the lead–sulfuric acid system. They show superior life characteristics and when employed in passenger cars outlive the car. In series connection, nickel-cadmium batteries must be designed against polarity reversal of a cell.

When cells are in series and discharged completely, small cell-capacity differences will cause one cell to reach complete discharge sooner than the others. The cell which reaches full discharge first will be driven into reverse by the others. When this happens oxygen will be produced at

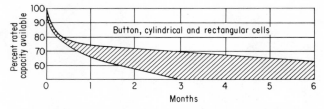

Fig. 119 Typical self-discharge curve of sealed nickel-cadmium cells in dry air. (*Eveready, Union Carbide Corporation.*)

the cadmium electrode and hydrogen at the nickel electrode. Gas pressure will increase as long as current is driven through the cell, and eventually it will either vent or burst. This condition is prevented in sealed nickel-cadmium cells by reducible material in the positive; also, nickel

TABLE 40 Sealed Nickel-Cadmium Cells

Volt-age	ANSI designation	Zinc-carbon equivalent (approx.)	Alkaline–manganese dioxide equivalent (approx.)	Current, ma	Service, mah	Number of cells and sizes	Diameter in.	Diameter mm	Length in.	Length mm	Width in.	Width mm	Height in.	Height mm	Weight, oz	Volume cu in.	Volume cu cm
1.25	K20			5	50	1/K20	0.914	23.21					0.252	6.40	0.12	0.06	0.94
1.25	K23			100	100	1/K23	0.914	23.21					0.272	6.91	0.33	0.15	2.34
1.25	K28			150	150	1/K28	0.914	23.21					0.355	9.02	0.40	0.16	2.50
1.25	K30			225	225		0.914	23.21					1.094	27.79	0.44	0.23	3.59
1.25	K32						0.563	14.30					0.212	5.38	0.56	0.27	4.21
1.25							1.365	34.67					0.583	14.81	0.50	0.31	4.84
1.25							0.914	23.21					0.315	8.00	0.66	0.36	5.62
1.25				400	400		0.595	15.11					0.968	24.59	1.17	0.61	9.52
1.25				450	450		0.562	14.27					1.969	50.01	0.75	0.43	6.71
1.25	K40	AA		450	450	1/K40	0.563	14.30					1.969	50.01	0.80	0.42	6.56
1.25	K45		L40	1,000	500	1/K45	1.256	31.90					1.980	50.29	1.0	0.43	6.71
1.25	K46	A		1,000	500		0.625	15.88					1.980	40.29	1.30	0.61	9.52
1.25	K47						1.031	26.19					1.031	26.19	0.78	0.87	13.58
1.25	K60			3,000	750	1/K60	0.900	22.86					1.684	42.77	1.7	0.96	14.99
1.25	K65			2,400	1,200	1/K65	0.900	22.86					1.666	42.32	1.7	0.96	14.99
1.25				1,000	1,000		0.862	21.89					0.394	10.01	2.0	0.85	13.27
1.25	K70	C	L70	1,000	1,000	1/K70	1.862	47.29					0.394	10.01	2.0	0.85	13.27
1.25	K75	1/2D		3,000	1,500	1/K75	1.631	41.42					1.805	45.85	2.3	1.35	21.09
1.25	K80		L80	2,000	2,000	1/K80	1.866	47.29					0.667	16.94	3.3	1.83	28.58
1.25	K85	D		4,600	2,300		1.344	34.13					1.508	38.30	3.2	1.82	28.25
1.25	K90		L90	2,000	2,000	1/K90	1.031	26.19					3.344	84.94	4.0	2.79	43.7
1.25	K95	F		3,000	6,000	1/K95	1.344	34.13					2.406	61.11	5.0	2.56	39.98
1.25				12,000	2,000		1.866	47.29					0.976	24.79	8.2	4.37	68.25
1.25				2,000	3,500		1.344	34.13					2.406	61.11	6.0	2.56	39.98
1.25				3,000	3,500		1.866	47.29					3.600	91.44	6.0	3.78	59.04
1.25				3,500	4,500		1.866	47.29	1.390	35.35	1.390	35.35	2.156	54.62	5.0	5.4	84.34
1.25				1,000	4,000		1.300	33.02	1.390	35.35	1.390	35.35	0.976	24.79	9.2	2.98	46.55
1.25				4,500	4,000				1.484	37.69	1.703	43.25	3.562	90.46	5.4	2.98	46.55
1.25				6,000	4,500				1.968	49.98	1.703	43.25	2.327	59.11	13.4	8.97	140.11
1.25				6,000	6,000				1.968	49.98	1.703	43.25	2.906	73.81	15.7	10.5	164.01
1.25				7,500	7,500								3.703	94.05	18.0	12.9	201.49
1.25	K100	G	L100	16,000	8,000	K100	1.344	34.13					4.015	101.98	9.8	5.25	82.00
1.25				11,000	11,000				3.593	91.26	1.06	29.92	4.158	105.61	28	16.5	257.73
1.25				15,000	15,000				3.593	91.26	1.390	35.30	4.468	113.48	35	19.8	309.27
1.25				15,000	15,000				3.593	91.26	1.703	43.26	4.468	113.48	42	25.8	402.99
1.25				19,000	19,000				3.593	91.26	1.015	25.78	4.468	113.48	50	29.7	463.91
1.25				23,000	23,000				1.531	38.88	1.531	38.88	3.828	97.23		8.97	140.11
6.25				900	500	5/C900			3.593	91.26	1.562	39.67	3.656	92.86	8.3	22.13	345.67
6.25					1,500	5/PP			3.875	100.96	1.562	39.67					
10.00				10,000	4,000	8/K90			6.406	162.71	3.093	78.56	2.406	61.11	61.0	26.15	408.46
15.00				1,000	500	12/45			2.810	71.37	2.810	71.38	1.320	33.53	12.5	10.42	162.76

hydroxide suppresses hydrogen evolution when the positive expires. With cadmium oxide it is possible to prevent hydrogen formation and to react the oxygen formed at the negative by the same basic process used to regulate pressure during overcharge.

A cell is considered protected against reversal of polarity if, after discharge at the 10-hr rate down to 1.1 v, it may receive an additional 5-hr discharge with the same current without being damaged or otherwise affected. Obviously the 1.25-v emf of the nickel-cadmium cell requires a greater number of units in a 6- or 12-v automotive battery than does a 2.0-v lead unit.

Sealed nickel-cadmium batteries are applied in alarm systems, emergency lighting, electric shavers, dictating machines, electric knives, toothbrushes, instruments, tape recorders, radios, transmitters and receivers, movie cameras, hearing aids, amplifiers, telemetering, as well as rechargeable cordless drills, tools, and appliances.

Table 40 gives the electrical performance, dimensions, and physical characteristics of typical sealed nickel-cadmium cells and batteries for the enumerated applications.

Silver-Cadmium Cells

The silver-cadmium cell has higher energy density per unit weight and per unit volume than the nickel-cadmium unit. This makes its application attractive where reduced size and weight are of importance. Silver-cadmium cells have an energy density of 1.5 wh/cu in. and about 14 wh/lb.

Silver-cadmium button cells are available in four sizes as shown in Table 41.

The silver-cadmium button cell utilizes pasted silver electrodes, sintered cadmium electrodes, and a three-ply separator, packaged in a metal container.

Silver-cadmium button cells have been life-cycled using a constant

TABLE 41 Silver-Cadmium Cells

Voltage	Gulton designation	Diameter		Thickness		Weight, oz	Cutoff voltage
		in.	mm	in.	mm		
1.4	AG .750	1.360	34.54	0.390	9.90	1.1	1.0
1.4	AG .500	1.360	34.54	0.240	6.09	0.69	1.0
1.4	AG .300	0.982	24.94	0.300	7.62	0.45	1.0
1.4	AG .120	0.893	22.68	0.230	5.84	0.28	1.0

current charge to a fixed voltage cutoff (1.52 v per cell) and discharged at constant current for a fixed time.

The charge-retention characteristics of silver-cadmium button cells are such that after a 30-day stand, cells retain in excess of 90% capacity.

Charging may be done using a constant current (C/10 or less) to a 1.52 v per cell cutoff, or using a constant potential mode (1.50 v per cell), limiting the initial surge current to C/2. Discharging may be done at either constant current or constant load. At up to moderate rates, it is recommended that the cell be discharged to 1.0 v cutoff. At higher rates, the cutoff voltage may be lowered to 0.8 v per cell.

chapter 16

Battery Charging: Theory and Practice

A battery charger is a source of direct current having a voltage higher than that of the fully charged battery. The positive terminal of the charging source is connected to the positive terminal of the battery so that the charging current flows through the battery in the direction opposite to that of the discharge current, as shown in Fig. 120.

The battery counter emf includes an emf characteristic of the electrode-electrolyte system and the combined polarization (voltages) of the two electrodes. The emf is relatively constant for the nickel-cadmium system but increases during the charging of the alkaline zinc–manganese dioxide system. The electrode polarization voltages increase during charge and are dependent on the state of charge and the magnitude of the charging current. The equation for the circuit of Fig. 120 is

$$E = E_c + IR_i \qquad (1)$$

Therefore the current flowing at any given time is

$$I = \frac{E - E_c}{R_i} \qquad (2)$$

Thus when the voltage of the battery E_c is equal to the charging source voltage E, no current will flow. When the voltage of the battery is less than that of the charging source current will flow into the battery and charge it, but if the battery voltage be higher, current will flow out of the battery and discharge it.

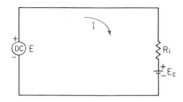

Fig. 120 Elementary charging circuit. E = Impressed voltage; I = Charging current; R_i = Internal battery resistance; E_c = Battery counter electromotive force. (*Eveready, Union Carbide Corporation.*)

The internal resistance R_i of most batteries is small, usually a fraction of an ohm. The change in the counter emf E_c during charge is relatively small. Large changes in charge current will result from small changes in the source voltage E. This may cause excessive currents which exceed the rates at which the batteries are capable of accepting. A method of reducing the current changes is to increase the source voltage E and add a series resistance to limit the current to the desired value.

In the secondary batteries, charge and discharge rates are related to capacities. The rate is the current which would completely discharge the battery in a specified number of hours. The current then becomes the ampere-hour capacity divided by the number of hours. Conversely, if the current be known, then the rate in hours is obtained by dividing capacity by the current.

There are three general methods of charging systems: (1) constant voltage, (2) constant current, and (3) taper current.

The constant-voltage method supplies a tapering current and limits the maximum battery voltage to a fixed value. At the start, the current for nickel-cadmium batteries may be as high as the 2-hr rate and drop to the 20-hr rate, or lower, toward the end of the charge. Consider a 4-ah nickel-cadmium cell having an internal resistance of 0.1 ohm being charged from a constant voltage of 1.45 v. At the start the counter emf is 1.25 v and at the end it is 1.43 v. Initially, the current and rate will be

Current $\qquad I = \dfrac{E - E_c}{R_i} = \dfrac{1.45 - 1.25}{0.1} = 2$ amp

Rate $\qquad \dfrac{\text{ah}}{I} = \dfrac{4}{2} = $ 2-hr rate and at the end

$$I = \frac{1.45 - 1.43}{0.1} = 0.2 \text{ amp} \qquad \frac{4}{0.2} = 20\text{-hr rate}$$

The voltage E must be correct and remain constant, requiring an expensive constant source of $\pm 1\%$. Constant voltage chargers for

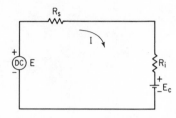

Fig. 121 Charging circuit with current limiting series resistor R_s. (*Eveready, Union Carbide Corporation.*)

nickel-cadmium batteries or alkaline manganese batteries are not recommended.

Constant current charging is cheapest and simplest. The supply voltage is much greater than the battery voltage and the current flow is limited with a large amount of series resistance (Fig. 121).

Equation (2) now becomes

$$I = \frac{E - E_c}{R_s + R_i} \qquad (3)$$

or

$$R_s = \frac{E - E_c}{I} - R_i \qquad (4)$$

If E be made large with respect to E_c the current becomes constant with small voltage changes. Assume that the same 4 ah nickel-cadmium cell is to be charged at the 10-hr rate of 0.40 amp from a 10-v source. The value for the series resistor can be calculated for the initial counter emf of 1.25 v and the internal resistance of 0.1 ohm:

$$R_s = \frac{10 - 1.25}{0.40} - 0.1 = 21.8 \text{ ohms}$$

The current at the end of charge when E_c has increased to 1.43 v will be

$$I = \frac{10 - 1.43}{21.8 + 0.1} = 0.39 \text{ amp}$$

Thus the current will decrease only by $2\frac{1}{2}\%$ during charge compared to a tenfold decrease for constant voltage charging.

Trickle charging is continuous constant current charging at low currents between the 300- and 3,000-hr rates.

In taper charging the current is high at the start and tapers off to a low current at the end. The taper results from the rise in voltage of the battery during charge. Taper charging can be used for alkaline manganese batteries which exhibit a voltage rise during charge. An inexpensive constant voltage ($\pm 3\%$ regulation) and a current-limiting resistor suffice. The maximum initial current should be the 4- or 6-hr rate.

Chargers include all components to form a system. The battery and charger are assumed to be at normal ambient temperatures of 70 to 80°F, and the charger "warm." Charging time refers to the interval required to recharge a 100% discharged battery. Most values are for a 100-ah cell.

Lead-acid and nickel-iron cell types have inherently different cell voltages, that of lead cells being nominally 2.0 v and that of nickel types 1.2 v. A nickel-iron unit requires a larger number of cells in the proportion of 5 to 3. A requirement of 36 v would use either an 18-cell lead battery or a 30-cell nickel type. The ampere-hour capacity would be the same.

Every charger has a slope characteristic or relationship between the output voltage and the current throughout its range. At any given voltage it delivers a given current and vice versa. The charger responds to the counter voltage of the battery connected to it.

This does not mean that any charger will charge any battery. If the charger voltage be less than the battery voltage, no current will flow in the charge direction. Conversely, if the charger voltage be sufficiently high that the current drawn is beyond its capabilities, the charger will overheat unless protected by a fuse.

Slope characteristics are illustrated in Fig. 122, and their resultant currents in Fig. 123. An infinite number of slopes are possible. Some

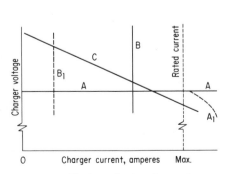

Fig. 122 Various basic slope characteristics. (*ESB Incorporated.*)

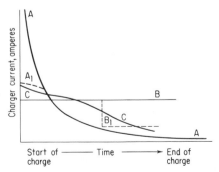

Fig. 123 Currents produced by basic slopes. (*ESB Incorporated.*)

chargers have "composite" slopes, with one used during the early charge and merging into another during the latter portion.

The significance of these slope lines is their angle. Their actual position, or value, varies with the application and is usually adjustable within a certain latitude.

Slope A, a "constant-voltage," or "constant-potential" type, represents a charger which maintains a constant voltage throughout its load range. This would, at nominal voltage values, deliver an extremely high current to a discharged battery, as shown by line A in Fig. 123. It has some form of current limitation such as shown at A_1 which causes the voltage to drop when the current exceeds rated load, preventing any further current increase.

This slope is used for the "float" operation of all types. At a selected voltage, the battery will accept a recharge if required, and otherwise remain in a fully charged state without appreciable overcharge.

Slope B, constant-current type, maintains the current at a constant

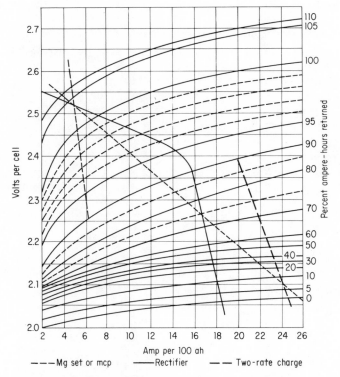

Fig. 124 Charge characteristics of lead-acid battery, with charger slope lines. (*ESB Incorporated.*)

value throughout the charge for the recharging of nickel-iron batteries, as they can accept a comparatively high current throughout the entire charge. It can be used for lead-acid types if sufficient time be available to charge throughout at the lower finish rate. It can also charge lead

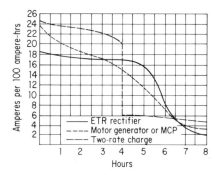

Fig. 125 Currents produced by typical chargers with slopes similar to those shown in Fig. 124. (*ESB Incorporated.*)

batteries in a shorter time by conducting the early part of the charge at a high rate and then shifting to a lower value (from B to B_1) to complete the charge.

Slope C is a "tapering" slope with high starting rate, gradually decreasing to a low finish rate applied with lead-acid batteries which require a complete recharge in 8 hr.

Figures 124 and 125 show several slopes on a lead battery characteristic and their corresponding current patterns. These separate characteristics can be obtained by test by a series of voltage readings taken at different current values and, in the case of the battery, at different states of charge. Both of course must be on a comparable basis, usually that of one 100-ah cell. By sequential increments of input, values of current, voltage, and time may be calculated.

For most nickel-cadmium battery applications the only available source of charging current is 117 to 120 v, 60 cycle ac. If such a source is con-

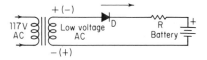

Fig. 126 Half-wave charging circuit using voltage step-down transformer and current-limiting resistor. (*Eveready, Union Carbide Corporation.*)

nected to a battery, the current would be charging the battery half of the time and discharging it during the other half, and there would be no net change in the charge in the battery. It is necessary to rectify the alternating current into direct current, that is, permit it to flow in one

182 Batteries and Energy Systems

direction only. Rectifiers act like check valves in a water line, permitting flow in one direction but blocking it in the other direction.

The rectification of ac may be either half wave or full wave. In half wave only one polarity of the ac is used as indicated in the circuit of Fig. 126.

Alkaline–manganese dioxide secondary batteries on charge exhibit a considerable voltage rise which can be related to their state of charge. Initially, battery voltage is about 1.3 v per cell and increases to about 1.70 to 1.75 v per cell as the battery approaches full charge.

Maximum charge rates are given in Table 42 for each cell size. These replace about 125% of the cell's rated capacity in 10 hr. Lower rates should be used where a longer charge period is available.

TABLE 42 Sizes and Capacity

Cell size	Rated ah capacity	Max. recommended charging current, amp
D	2.0	0.25
F	3.2	0.40
G	4.0	0.50

chapter 17

Regenerative Electrochemical Systems

The regenerative emf systems emphasize the conversion of heat into electrical energy. These are the thermally regenerative systems. However, there are photochemical, chemical, and electrical regenerative emf systems of interest. The latter are secondary batteries which could be used in electrical propulsion. The emf systems suitable to thermal regeneration are much more restricted in number than systems suitable to electrical regeneration. The bimetallic cells show promise for both thermal and electrical regeneration.

Energy conversion systems which have received the most research and development attention in the past 10 years are metal vapor turbines, thermionic diodes, magnetohydrodynamic generators, and thermoelectric devices.

The regenerative electrochemical converter can be virtually free of mechanical parts and, by comparison with the other systems, the flowing liquids or gases are low-velocity streams.

Liebhafsky[1] says,

> Among the increasingly complex and sophisticated methods of energy conversion needed today, regenerative electrochemical systems occupy a

place. In this introduction, the author attempts to mitigate the confusion that exists in the naming of these systems, to discuss their thermodynamics on the basis of simple examples, and to show that these systems may be regarded as foreshadowed by the work of Grove (1839). If experience with fuel batteries be a valid guide, the development of practical regenerative electrochemical systems will encounter many difficult engineering problems, and the difficulty of developing such systems will increase with the complexity of the transport problems involved.

Angus[2] described a

gas concentration cell as a means of converting thermal to electrical energy. Initial results using I_2 vapor and a PhI_2 electrolyte are given, as well as estimated characteristics of cells using alkali metal vapors and alkali metal halide electrolytes.

Berger and Strier[3] state that

Sintered zirconium phosphate membranes containing zeolites have significant water absorptive capacities over a temperature range of ambient to 150°C. They are sufficiently conductive for fuel cell applications over this temperature range. Such membranes have transverse strengths of 5,000 to 6,000 psi. They readily gain and lose water vapor in a reversible manner while maintaining good stability.

Herédy, Iverson, Ulrich, and Recht[4]

studied the chemical and electrochemical characteristics of sodium amalgam galvanic cells. A static electrode cell, contained in a stainless steel pressure vessel, was operated at temperatures from 477 to 510°C under 140 to 180 psig argon cover gas pressure. A molten eutectic mixture of sodium salts was the electrolyte. Current densities as high as 200 ma/sq cm were achieved without appreciable electrode polarization. A flowing electrode cell with a tubular electrode matrix was designed, built, and tested. The anode and the cathode compartments were supplied with continuous streams of concentrated and dilute amalgam, respectively. A complete thermally regenerative system (a flowing electrode cell coupled with a regeneration loop) was operated for a period of 1,200 hr at a cell temperature of about 490°C.

Groce and Oldenkamp[5] studied the design of a thermally regenerative sodium-mercury galvanic system. They found that

in operating Atomics International's Thermally Regenerative Alloy Cell (TRAC) system, the sodium amalgam stream from the cell battery must be converted to a sodium-rich amalgam and mercury. This is done in the regenerator which distills and separates a mercury fraction, condenses the mercury vapor, cools both streams, and then recirculates them to the cells. A test loop was built and operated to study these regeneration processes. Two tests were made with the regeneration loop connected to single-matrix TRAC cells—one for 116 hr, another for 1,197 hr. The cell internal resistance did not change during either test, indicating that the cell materials are compatible with the working fluids under flow conditions. The tests also demonstrated the long-term operability of the TRAC system.

McCully, Rymarz, and Nicholson[6] studied regenerative chloride systems for conversion of heat to electrical energy.

Closed cycle combinations of thermochemical and electrochemical reactions provide a potentially simple and efficient mechanism for conversion of heat to electrical energy. Several chloride systems meet the primary criteria for this cyclic process. Molten tellurium dichloride is ionic and a suitable anode material but is gaseous at the regeneration temperature complicating the separation from chlorine in this step. Antimony trichloride is less ideal as an anode but is easily separated in the regeneration step. Successful operation of the cyclic process in a single device has provided potentials from 0.3 to 0.5 v. Projected operating efficiencies range to 28% of the accepted heat.

Anderson, Greenberg, and Adams[7]

applied expressions derived for assessing theoretical performance characteristics to regenerative emf cells with molten salt electrolytes. Predicted efficiencies for systems based on metal iodides range from 15 to 30%, with power densities of 5 to 140 milliwatts/sq cm (mwsc). Rapid recombination rates severely limit systems regenerable by thermal dissociation, and regeneration by high-temperature electrolysis and by thermal dissociation are thermodynamically equivalent processes. Nonisothermal processes suggest that there is little evidence for ionic entropy of transport effects in molten salt electrolytes. For a photoregenerative system, observed conversion efficiencies are below 1% and agree with predictions based on a theoretical model.

Photochemically regenerative electrochemical devices must compete with solar cell–storage battery systems. Therefore, efficiencies must be comparable or storage capacity must compensate for lower efficiencies. The photochemical system proved to have operational efficiencies two orders of magnitude lower than desired. In this case, the energy storage capacity does not compensate for the low efficiency. Many of the factors which set an upper limit on the efficiency of solar cells, however, cannot be applied to photoregenerative chemical systems.

Fischer[8] discussed laboratory studies of intermetallic cells. He states

General relationships useful in selecting and operating a bimetallic regenerative cell may be deduced from the phase diagram of the system. The equilibrium pressure of the cell system is effectively fixed by the condensation and temperature in the regenerator. This pressure will fix the applicable temperature-composition phase diagram and will determine whether there is the necessary separation of the liquid-vapor loop from the liquid-solid regions. The composition of material to be returned to the cathode will dictate the regeneration temperature. The melting point of any compound formed should not be so high that separating the vapor-liquid and liquid-solid regions cannot be achieved at a practical operating pressure. Additional pressure conditions may be imposed if azeotropy is present and is to be eliminated.

Matsen[9] states that

chemically regenerative, fuel cell systems have long been studied in the hope of circumventing the problems of direct electrochemical reaction of fuels and air. These cells employ reaction intermediates which readily react at the electrodes and which are then regenerated with fuel and/or air. Such schemes received considerable attention at the turn of the century, and some studies have been made recently with difficult fuels such as coal. The trend, however, is to highly reactive fuels which readily undergo direct electrode reaction and to military or space applications where the additional weight and complexity of chemically regenerative systems is highly undesirable.

Findl and Klein[10] studied a H_2-O_2 fuel cell storage battery.

The regenerative H_2-O_2 cell is basically a combination of a H_2-O_2 primary fuel cell and a water electrolysis cell in one compact package. Various-size experimental units have been built from single cells (6-in.-diameter electrodes) to a 34-cell series unit with nominal ratings of 28 v, 500 w, and 21-ah capacity. Electrical testing has been conducted over a range of variables. Cycle life performance of 350 cycles on a test regime of 35 min discharge and 65 min charge has been achieved.

Unlike conventional batteries, charge rates of the H_2-O_2 cell can be as rapid as desired, limited in practice only by the heat rejection capability of the system and charge-power, equipment limitations. Where conventional battery charge rates must be kept below the point at which the water electrolyte decomposes, the H_2-O_2 cell operates by water decomposition. During discharge, the H_2-O_2 fuel-cell mode of operation is capable of sustaining 2 to 3 times the current density of standard batteries with little change in performance.

A critical requirement of all batteries is the energy density they are capable of producing. The literature contains numbers ranging from 0.5 to 100 wh/lb for conventional batteries. For practical space applications as secondary batteries, the useful watt hr/lb values are 1 to 2 for Ni-Cd cells, 5 to 8 for Ag-Cd cells, and 8 to 12 for Ag-Zn cells.

REFERENCES

1. H. A. Liebhafsky, Introduction to Regenerative Cells, in Robert F. Gould (ed.), "Regenerative Emf Cells," Advances in Chemistry Series, pp. 1–10, ACS, Washington, D.C., 1967.
2. John C. Angus, Continuous Gas Concentration Cells as Thermally Regenerative, Galvanic Cells, *ibid.*, pp. 11–16.

3. C. Berger and M. P. Strier, Solid Inorganic Electrolyte Regenerative Fuel Cell System, *ibid.*, pp. 17–29.
4. L. A. Herédy, M. L. Iverson, G. D. Ulrich, and H. L. Recht, Development of a Thermally Regenerative Sodium-Mercury Galvanic System, Part I, Electrochemical and Chemical Behavior of Sodium-Mercury Galvanic Cells, *ibid.*, pp. 30–42.
5. I. J. Groce and R. D. Oldenkamp, Development of a Thermally Regenerative Sodium-Mercury Galvanic System, Part II, Design, Construction, and Testing of a Thermally Regenerative Sodium-Mercury Galvanic System, *ibid.*, pp. 43–52.
6. C. R. McCully, T. M. Rymarz, and S. B. Nicholson, Regenerative Chloride Systems for Conversion of Heat to Electrical Energy, *ibid.*, pp. 198–212.
7. L. B. Anderson, S. A. Greenberg, and G. B. Adams, Thermally and Photochemically Regenerative Electrochemical Systems, *ibid.*, pp. 213–275.
8. A. K. Fischer, Phase Diagram Considerations for the Regenerative Bimetallic Cell, *ibid.*, pp. 121–135.
9. J. M. Matsen, Chemically Regenerative Fuel Cell Systems, *ibid.*, pp. 277–291.
10. E. Findl and M. Klein, Electrolytically Regenerative Hydrogen/Oxygen Fuel Cell Battery, *ibid.*, 292–305.

chapter 18
Solar Cells and Related Systems

The principle of the solar cell[1] is shown in Fig. 127. When light strikes a semiconductor, an electron may be dislodged from its normal position or circuit, leaving a positively charged hole. Both electron and hole will then be available for the conduction of electricity if they can be prevented from recombining and thus neutralizing each other. The p-n junction provides a built-in electric field that pulls the electrons into the n or negative side of the junction and the holes into the p or positive side before many of them recombine.

The solar cell is a large p-n junction oriented to face the sun, and the top layer is made very thin so that as many as possible of the effective photons may penetrate to the vicinity of the junction.

With silicon, the effective parts of solar energy are absorbed in the outer 0.001-in. layer, and to be collected the electron-hole pairs must be produced within about 0.0001 in. of the junction. The surface must have a high conductivity.

One method of forming the p layer is to heat a plate of n-type silicon (arsenic-doped) in a gas containing boron. The boron diffuses into the silicon to a depth determined by time and temperature. At the surface, the silicon is very heavily "doped" with boron. Further into the crystal the concentration decreases until, at about 0.0001 in., it gets so low that the n-type arsenic-doped silicon of the crystal predominates. This change in conductivity type defines the position of the junction. The

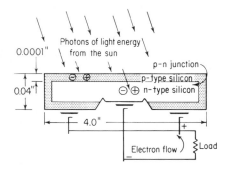

Fig. 127 Principle of the solar battery.

p layer is then removed from a portion at the bottom surface and contact is made to the body of the crystal, as shown in the center contact of Fig. 127. Contact to the p layer then completes the electrical arrangements.

The energy efficiency of conversion of sunlight to power of the cells was 6 to 11%. Five cells connected in series with a total area of about 10 sq cm charged a 1.35-v storage battery. On a clear day the charging current averages about 25 ma from 7 A.M. to 5 P.M.

Solar batteries made of p-type and n-type silicon provide power for satellite instrumentation. In an earth satellite such cells generate electricity about 60% of the time (the remainder being that portion of the orbit in the shade of the earth), and during that time they not only furnish power for the instruments and radio but also charge small nickel-cadmium batteries which take over during the "night time."

The photoelectric tube is based on the property of metals, such as sodium, potassium, rubidium, calcium, cadmium, selenium, or cerium, which give off electrons when they are exposed to ordinary light. A photoelectric tube may be constructed by coating a portion of the inside of an evacuated glass bulb with a layer of the metal and connecting this as one terminal of the circuit, with a metal electrode placed inside the bulb near the coated surface as the other terminal. As long as light falls on the photosensitive metal layer, a current will flow, the strength of which will be dependent upon the intensity of the light. The current is very small and may be amplified by a three-element vacuum tube as shown in Fig. 128. The metallic coating on the glass may be replaced by a plate of the photosensitive metal. High-vacuum photoelectric tubes have a high voltage drop and require a high polarizing potential.

190 Batteries and Energy Systems

This can be reduced if gas at a low pressure be used in the tube, with a decrease, however, of permanence of the characteristics of the unit. Both types of tube are employed for television for the transmission of photographs or copy by radio or wire lines. They find application for inspection, for sorting of products which vary in color, for the detection of

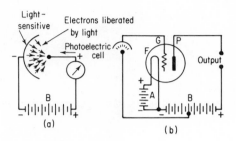

Fig. 128 Amplification of photoelectric tube current.

smoke, as guards on machines where a continuous ray of light is directed on the tube and any object which cuts off the light makes the machine inoperative, as well as for other purposes where a variation of light is to be detected.

Photovoltaic Cells

Photovoltaic cells are devices containing two electrodes immersed in a suitable electrolyte, between which electrodes an emf is produced or a change of emf effected when electromagnetic radiation such as light is incident on the photoactive elements of the cells. An example is a cell of $Cu|Cu_2O$ as one electrode, $Pb(NO_3)_2$ as the electrolyte, and Pb as the other electrode, having a sensitivity of 150 μa/lumen.

Electronic tubes may be used as potentiometers, with the glass indicator electrodes for acidity control where the hydrogen, quinhydrone, and metallic electrodes cannot be used because of rapid poisoning.

REFERENCE

1. *Bell Labs. Record*, July, 1955.

chapter 19

Development and Specialized Application Cells

The lithium–nickel halide battery represents a potential high-energy density source of electric power. It has reached the stage of development where laboratory experimental models in sizes up to 20 ah have been built and tested successfully.

Lithium is the earth's lightest solid element. Weighing only about seven times as much as hydrogen (the lightest element in the universe), lithium also has one of the highest electrochemical potentials of any feasible anode material.

Compared with lead, lithium has enormous theoretical advantages for batteries suitable for propelling an electric car. While lithium weighs only about one-thirtieth as much as an equal volume of lead, it can generate up to eight times as much electricity when coupled with a suitable cathode in the presence of an effective electrolyte.

For example, if a suitable electrolyte to conduct current between the two electrodes is furnished, a 1-lb electrochemical couple consisting of a lithium anode and an elemental fluorine cathode could deliver a theoretical maximum of 2,770 wh of electricity. By comparison, the theoretical limit of the familiar lead-acid battery is only about 80 wh/lb.

As with the antique electrics introduced before the turn of the century and the recently converted Renaults, the crux of the problem with both the GM Electrovair and the Ford "city car" is the batteries. Spurred not only by the pressure for an electric car but also by the nation's space exploration program, more than a dozen companies are conducting studies, sponsored by NASA, the Army, the Navy, and the Air Force, to develop more efficient batteries.

Major attention in these efforts is directed toward rechargeable or secondary batteries rather than the throwaway primary types. Firms doing development work on rechargeable lithium batteries include Electrochimica, Exide (ESB), General Motors, Globe-Union, General Atomics, Honeywell, Inc., Lockheed, Mallory, Rocketdyne, Tyco Laboratories, Union Carbide, and Whitaker Corporation.

Instead of using a solid lithium electrode, Gulton devised a "dispersed" lithium anode. A mineral oil dispersion of tiny lithium particles, each about 0.1 mm in diameter, is dispersed with nickel flakes, monel flakes, or graphite, all of which are good conductors. The mixture is then bonded to a metal grid of nickel, monel, or inconel by a binder such as polyethylene and carboxymethylcellulose.

To prevent oxidation that would damage the lithium during preparation of the mixture, Gulton keeps the materials under a dry, inert atmosphere of argon. After the lithium dispersion is pressed into the metal grid, the lithium electrode is hermetically sealed in the cell.

The Gulton battery, weighing 250 lb, would provide enough energy to propel a compact automobile for 150 miles without recharging. Research is currently underway to develop a lighter-weight battery.

Because lithium reacts vigorously with water, a nonaqueous solvent is used for the electrolyte. Among the organic liquids found useful to date are propylene carbonate and dimethyl sulfoxide. Potassium hexafluorophosphate and aluminum chloride have been used as conductors, and ion flow has been provided by lithium chloride and lithium fluoride.

The prismatic lithium–nickel fluoride cells are assembled in sealed, molded, polypropylene containers, with solid terminals extending beyond the cover for intercell connection into batteries.

The lithium battery has a theoretical capability of 600 wh/lb, though the prototype has achieved only 100 wh/lb. This tops nickel-cadmium (26 wh), silver-cadmium (35 wh), and silver-zinc (75 wh).

After 30 days of charged stand, at ambient room temperature, cells can be expected to deliver over 95% of their rated capacities.

Charging of the lithium–nickel fluoride cell may be accomplished by using a constant current source with limiting voltage to cut off at 3.4 v per cell, or constant potential mode (3.4 v per cell), limiting the surge charge current to the 2-hr rate (C/2).

Lithium cells are normally furnished completely sealed and fully charged.

GM's Allison division has developed its own concept of a lithium battery for propelling an electric automobile. Lithium would be used both as an electrode and as a component in the current-carrying electrolyte.

During discharge of the Allison cell, lithium moves by capillary action from a lithium storage compartment, where it floats on lithium chloride, to a reaction zone. Chlorine is fed from an external storage cell to a centrally located graphite electrode, the outer shell of which is porous. The reaction produces an excellent flow of electricity and makes more electrolyte by forming lithium chloride.

To charge the Allison cell, an external electric current is applied to the lithium chloride formed when the cell is working. The applied current causes the lithium and chlorine to separate; the lithium rising into its storage region and the chlorine being carried out of the cell to be reprocessed and stored.

Allison indicates that the lithium–lithium chloride battery, while strictly experimental, shows very good power density and excellent discharge and recharge rates. If the concept proves out, it could meet the heavy peak power demands that will be placed on the battery in any electric car when passing, accelerating rapidly, climbing hills, etc.

Fast recharge rates will permit the Allison battery to be brought back to its rated charge quickly, perhaps during a stop at a service station of the future equipped with commercial chargers instead of gas pumps.

Johnson and Heinrich[1] conducted investigations on the lithium hydride, regenerative emf cell and demonstrated the feasibility of a hydrogen-permeable membrane as the cathode on a cell. Emf techniques have been used to determine the fundamental thermodynamic quantities (ΔG_f^0, ΔH_f^0, ΔH_f^0) for the lithium hydride cell product. Temperature-composition data define electrolytes suitable for the lithium hydride cell.

At Argonne National Laboratory, a lithium hydride cell with a vanadium diaphragm was operated for 540 hr at 525°C, giving current densities that exceeded 200 ma/sq cm at one-half open-circuit voltage (0.3 v). Even though this current density is only about one-half the theoretical current density (for a 0.010 in.-thick vanadium diaphragm at 500°C), it was achieved without any high-temperature pretreatment of the metal diaphragm. No mechanical problems arose during the cell run. The slope of a plot of cell emf measured against the logarithm of the hydrogen pressure corresponded to the expected Nernst slope of 1, indicating the reversibility of the H_2, H^- couple at the vanadium electrode.

The electrolyte or the cathode construction, for the most part, determines design parameters. Because lithium metal has unique characteristics, the anode design incorporating either a sintered-metal or controlled-porosity disk will be most suitable. If one uses the LiH|LiCl|LiF electrolyte system (minimum at 443°C), a permeable membrane for the cathode would be a possible choice.

The advantage given by the permeable membrane is its ready solution to the assorted problems that have long plagued porous, metal-gas electrodes. As the permeability studies indicate, however, the choice of metal for this application is limited to iron and vanadium. Iron has a low hydrogen permeation rate, and vanadium requires pretreatment before it is suitable. It should be emphasized that in a permeable membrane, the temperature of the cell operation must be above about 350°C (considering a 0.002-in.-thick vanadium diaphragm). Only at this temperature and above can the hydrogen permeation rates through the cathode metal support reasonable current densities.

In choosing the LiH|LiCl|LiI electrolyte system (ternary eutectic melting at 325°C), one gains a number of advantages because the cell can now operate at lower temperatures. Cell voltages will be significantly higher. This electrolyte is at or near saturation with respect to the lithium hydride component. At saturation, the cell voltage is 0.45 v. Regeneration pressures can be significantly increased by sending to the regenerator a mixture very rich in lithium hydride (90 to 95 mole % LiH). Such a mixture can be obtained by precipitating lithium hydride as a pure component from this electrolyte near cell-operating temperatures.

Eliason, Adams, and Kennedy[2] reported on a lithium-chlorine cell.

A fuel cell—i.e., fuel storage external to the power module based on the reactions:

Anode: $Li \rightarrow Li^+ + e^-$
Cathode: $\frac{1}{2}Cl_2 + e^- \rightarrow Cl^-$
Total: $Li(l) + \frac{1}{2}Cl_2(g) \rightarrow LiCl$ (LiCl electrolyte, 625°C)

was constructed and delivered 650 w (370 amp at 1.75 v). Power density was 8,000 w/sq ft at maximum power. The cell was self-sustaining under load and, in fact, required cooling when operated at maximum power. Principles of secondary operation of lithium|chlorine batteries have been developed, and laboratory investigation is in progress. The design features necessary to build a cell which does not fail in the presence of these energetic and high corrosive reactants are given.

Fortunately, lithium chloride does not wet graphite, and fairly large pressures (0.1 to 0.6 atm typical) are required to flood the porous graphite cathode. The same problems exist at the lithium electrode, and again a pressure differential must be established to force the liquid metal into the anode chamber. Gravity can be used in some cell designs because lithium floats on molten lithium chloride.

Lithium chloride is not a good medium for electrolysis. Lithium has a definite solubility (0.5 to 1 mole %) in its own salts, and because the chlorine is in close proximity the resulting current efficiency is poor. LiCl-KCl eutectic allows high efficiency operation at a lower temperature (approximately 400°C) and still produces pure lithium metal. Even with 1- to 2-in. electrode spacing, the operating cell voltage is less than 5 v. It is reasonable to assume that by decreasing electrode spacing to 0.5 in. or less (as has been done in the discharge cycle), the operating voltage can be reduced to 4.0 to 4.5 v. A battery discharging at 2.8 v and charging at 4.2 v would have a voltage efficiency of 67%. If the current efficiency were 90%, an overall efficiency of 60% would be achieved. Experiments have shown that these efficiencies can be realized in the laboratory provided the current density during charge is kept low (less than 0.5 asc). The lithium chloride product from discharge would flow into a cell containing sufficient KCl to form a low melting mixture. In the charging cycle, lithium could be deposited and pumped to the discharge cycle reservoir. Chlorine would be collected at the top of the cell and compressed in a tank for future delivery in the discharge cycle. With two sets of electrodes, optimum configuration and active areas for the two operations could be used.

Thermal batteries are based upon electrolytes of inorganic salts that are solid and nonconducting at normal missile and space vehicle operating temperatures. Heated, the electrolyte melts and becomes conductive. The battery functions as a reduction-oxidation process. The battery remains functional with a slow but constant increase in internal impedance with respect to time. The loss of heat from the cell and the accumulation of the waste products of the reaction increase internal impedance.

Combinations of calcium or magnesium anodes with reducible cathodes produce a variety of voltages and capacities. A typical system is calcium|lithium chloride|potassium chloride|calcium chromate. Calcium metal serves as the negative electrode. The positive is a saturated calcium chromate. The collectors and intercell connectors are iron, nickel, or an alloy. The inorganic electrolyte is potassium chloride and lithium chloride in the eutectic formulation. A nongaseous exothermic solid fuel heats the cells.

Thermal cells are constructed either as a pellet type or as a closed cup type, shown in Figs. 129 and 130. The pellet is made by mixing an inorganic binder which acts as a separator with electrolyte eutectic, and compacting them into a pellet. Finely ground positive active material (e.g., calcium chromate) may either be mixed in with the original electrolyte binder or may be the basis for another pellet. The pellet cell may be a two-layer type (Fig. 129) or may be single layer.

The electrolyte and separator for the closed cup cell are impregnated glass tape with the electrolyte salts cut to fit the cell. The active positive material is mixed with glass and quartz fibers. This is formed as a thin paper cut to fit the cell. The parts are placed in the cell cup. As shown in Fig. 130, the closed cup is two cells in parallel with a common negative in the center. The internal heat can be adjusted to allow the ambient temperature to vary by as much as 230°F with only slight degradation of the discharge capability.

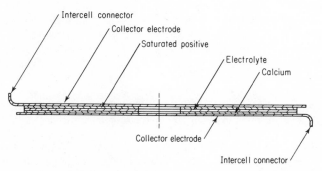

Fig. 129 Cross section of a typical pellet-type cell. (*Eagle-Picher Industries, Inc.*)

For mechanical activation, the primer is installed in the center of either end of the battery. A 20-in.-oz force applied to the center of the primer by a 0.02 radius spherical point is required for detonation. Electrical initiation is by two redundant matches, requiring an input of 580

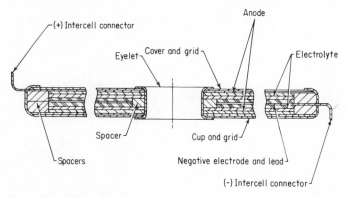

Fig. 130 Cross section of a typical cup-type cell. (*Eagle-Picher Industries, Inc.*)

ma for a minimum of 20 msec for ignition. The nominal resistance is 4.5 ± 0.5 ohms per match.

Cells are stacked into a cylindrical configuration with the diameter of the cells and the number in the stack being varied to provide various capacities and voltages. The stack or stacks of cells are insulated electrically and thermally and encased in a hermetically sealed metal container.

The usable voltage of thermal batteries ranges from 3 v per cell to a fraction of a volt, depending upon the discharge characteristics required and the electrochemical systems selected. Ordinarily under a constant resistive load, the normal voltage tolerance required for a thermal battery is ±10% of the nominal. They may be designed to produce several hundred amperes for a short period of time if a low voltage per cell can be tolerated.

The advantages of the thermal battery system are (1) operational temperature range capability of −65 to 200°F, (2) a weight-to-volume ratio of approximately 1.4 oz/cu in., (3) shelf and storage life, (4) versatility of electrical output, (5) fast activation, and (6) remote activation, either electrical or mechanical.

The disadvantage is the shortness of life. The life is limited to the length of time that the heat may be retained in the cell sufficient to keep the electrolyte in the molten condition. The majority of the applications are of less than 2 min total useful life. Since total life is dependent upon heat retention, the batteries of longer life are normally larger and heavier to allow for greater heat capacity and insulation.

Coleman[3] described radioisotopic, high-potential low-current sources. Beta rays consist of a stream of high-energy electrons. A rugged cell may be made by coating the isotope such as strontium 90 between two strips of gold foil, thus forming an emitter electrode and collecting the

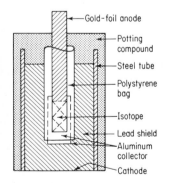

Fig. 131 Radioisotope battery.

electrons on an outer metallic container of aluminum, as in Fig. 131. Such cells are capable of producing a high voltage if time elapses for charges to build up. With a large capacitor in parallel, a large enough charge may be built up to operate one-shot devices, giving output currents of 40 μa at zero voltage and lower currents at maximum voltages of 6,000 v after 2 months.

Patterson, Moos Research has developed a family of nuclear batteries that convert nuclear energy directly into electrical energy. These batteries are small, rugged, lightweight, extremely reliable, and have an operational and shelf life of about 25 years. The batteries weigh 9 g and are $9/16$ in. long $\times$ $1/2$ in. diameter. They can be supplied in current ratings from 10 to 200 μμa and have a linear charging rate of 500 v.

Battery construction is simply an emitter electrode coated with the radioactive isotope (β-emitter) and separated from a brass collector case by a solid dielectric. Under AEC (Atomic Energy Commission) regulations, the battery may be treated as a conventional sealed source. Energy from radioactive isotopes is available for predictable and long periods (in this case 25 years) and is independent of adverse physical and chemical conditions. Therefore, nuclear batteries will not be affected by short circuiting, pressure, temperature, vibration, or brutal accelerating forces.

These are employed for timing circuits, biasing elements, control circuits, transistor power supplies, constant current or voltage reference sources, polarizing elements, and power supplies for radiation measurement instrumentation. Characteristics are given in Table 43.

Raypaks are a family of hermetically sealed nuclear power packs for applications requiring: (1) reliable pulse energies from 800 to 337,000 ergs, (2) regulated voltages at specific voltages between 75 and 750 v, (3) long shelf and use life (over 25 years), (4) miniature size and minimum weight, (5) temperature and pressure independence over wide ranges, and (6) ability to withstand brutal accelerating forces.

Raypaks are made up of a nuclear battery in conjunction with a capacitor and voltage regulation network. These power packs are supplied in a hermetically sealed metal case in which the electrical components are completely encapsulated. The output of the Raypak is provided at a standard connector. Because of its construction and makeup, a Raypak will not be affected by extremes in temperature, pressure, vibration, or

TABLE 43 Characteristics of Nuclear Batteries

Equilibrium voltage (70°F, 20% relative humidity)	Current, μμa ($\pm 10\%$)	Capacitance, μμf
5,000–10,000	10	4
5,000–10,000	100	4
5,000–10,000	200	4

TABLE 44 Characteristics of Raypaks

Voltage	Output energy, ergs	
	Single	Split
375	42,000	
375	....	20,000–20,000
750	337,000	
750	....	165,000–165,000

acceleration, and intermittent or continuous short-circuiting will not affect its shelf or use life.

There are sources of pulse energies from 800 to 337,000 ergs for trigger power in weapon and ordnance systems. For all stored power applications, space and weight must be kept to a minimum, long shelf life and use life are required, and ability to withstand wide temperature and pressure changes including brutal accelerating forces is mandatory. Characteristics are given in Table 44.

Betachrons are a family of hermetically sealed nuclear powered delay timers with the following properties: (1) complete choice of actuating time from milliseconds to 40 hr, (2) accuracy of $\pm 3\%$ over temperature range of $-65°$ to $+165°F$, (3) a self-contained nuclear power source which provides 25-year shelf and/or use life, (4) miniature size, $2\frac{1}{2}$ in. diameter $\times$ $1\frac{7}{8}$ in. length, weight 0.4 lb, (5) complete testability before use for added proof of reliability of timer and energy transfer, and (6) ability to withstand extremes in pressure, vibration, jolt, jumble, and brutal accelerating forces.

Betachrons are made up of a nuclear battery with associated circuitry which provides for fixed electrical time delays. Energy output is provided at a standard connector. Initiation of delay timer can be accomplished by snap action switch, pull wire, or electrical signal. Betachrons are applicable to pilot ejection systems, missile arming and safety systems, missile self-destructors, missile parachute recovery, warheads, satellite timers, and alarm systems, e.g., remote area pipeline and shutoff actuators. Characteristics are given in Table 45.

TABLE 45 Characteristics of Betachrons

Time range	Output energy, ergs
1 ms–60 sec	10,000
1 min–60 min	10,000
1 hr–40 hr	10,000

REFERENCES

1. C. E. Johnson and R. R. Heinrich, Thermodynamics of the Lithium Hydride Regenerative Cell, in Robert F. Gould (ed.), "Regenerative Emf Cells," Advances in Chemistry Series, pp. 105–120, ACS, Washington, D.C., 1967.
2. R. Eliason, J. Adams, and J. Kennedy, "Regenerative Emf Cells," Advances in Chemistry Series, Robert F. Gould, editor, pp. 186–197, ACS, Washington, D.C., 1967.
3. J. H. Coleman, *Nucleonics*, 11(12): 42–45 (1953).

chapter 20

Electric Cars and Batteries

There is a growing clamor over air pollution, automotive safety, noise, traffic congestion, more parking space, and more freeways. Indeed, some of these problems have become sufficiently acute as to warrant restrictive state and federal legislation.

Nonetheless, the automobile is now a necessity to millions of people—at least while no obviously eligible competitor is in sight. The motorist's past devotion to the gasoline-powered car cannot be counted upon indefinitely. A displaced suitor of the past clamors for attention. The contender is the electric car.

Because casings, connectors, and other accessories are needed, no electrochemical system can be built into a practical battery that will deliver anywhere near the theoretical potential.

Historically, the limited range and low speed of the electrics powered by lead-acid and alkaline batteries led to their replacement by the internal-combustion engine. Recently built electric cars perform little better than the old.

Edison took 10 years to perfect his nickel-iron alkaline storage battery. Storage batteries had been available ever since the mid-1850s when

Gaston Planté successfully produced the first practical secondary or rechargeable battery cell. The storage battery remained, however, little more than a laboratory curiosity for a decade or so. Then, during the 1870 to 1880 period, batteries commanded attention as a result of development of more efficient dynamos or generators and the rise of the electric power industry, which followed the success of Thomas Edison's Pearl Street generating station in New York City, opened in 1882.

Better generators now meant that storage batteries could be formed (the process by which plates are electrochemically activated) and charged more quickly and economically. Previously this was a costly and time-consuming process done only through primary batteries.

The demand for storage batteries grew with the development of electric power systems. These early dc systems, limited generally to metropolitan areas, supplied power for incandescent lamps which were replacing gas for home and street lighting. The generators did not supply power directly to customers, but rather to banks of storage batteries at the generating stations. It was these batteries which in turn furnished the power. Then, during daylight hours, the generators would recharge the batteries.

In the 1880's the electric automobile and the electric street truck added to the growing demand for storage batteries. True, there was a spate of "gas buggies" and "steamers" after about 1875. The early gas buggies, however, were noisy, hot, and smelly; they were also quite unreliable. The steamers were quieter, but they were also unpredictable.

The early electric vehicles had their problems, too. The batteries in these automobiles were the lead-acid type. They were not only heavy, but also highly inefficient—delivering only 4 to 6 wh/lb of battery—they were inherently self-destructive, and they were very short-lived. Replacement of batteries in electric "horseless carriages" was often required only a few months after the batteries had been put into service.

Edison's first venture with his new type of storage battery, however, was a failure. The early cells hardly equaled the claims made for them. It wasn't until 1910 that Edison was again ready to resume production of his storage battery with the almost accidental discovery that by the addition of some lithium to the potash electrolyte, the electrical capacity of cells became dependably more consistent. Edison himself never could quite explain why cell capacity of his nickel-iron battery could be improved by adding lithium.

Edison's new nickel-iron battery of 1910 was successful. Thousands of A cells were produced that year for manufacturers and owners of electric autos and street trucks.

The Edison battery enjoyed popularity in electric street vehicles before and after World War I. Internal-combustion engines increased the range and performance of gasoline vehicles. Dairy, bakery, and department store owners preferred electric street trucks because of their efficiency and economy in the stop-and-go operation of city traffic.

The electric vehicle market for the nickel-iron battery slowly disappeared. The "miner's safety cap lamp" utilized a small, compact version of a nickel-iron battery, worn on the belt to provide dependable, hazard-free illumination underground.

Railroads began using alkaline batteries widely for lighting, air conditioning railway passenger cars, powering controls, and emergency lights in subway and rapid transit cars. Service lives of up to 25 and in some cases 40 years have not been unusual in railway signal service.

In low-lift walkies and hand tractors, it has not been unusual for the nickel-iron battery to deliver 15 or even 20 years of service—practically for the life of the truck.

In the 1960 period the Henney Kilowatt is a converted Renault sedan in which the engine and drive train have been replaced with electric motors, controls, and 900 lb of lead-acid batteries. The Yardney Electric Company model was a converted Renault Dauphine but uses 300 lb of silver-zinc batteries, developed originally for aircraft use. Having less dead weight in its batteries, the Yardney is peppier than the Kilowatt; but neither performs much better than the electrics of yesteryear.

A Henney delivers 43 miles an hour (mph) on level ground and can run about 35 miles before the batteries must be recharged. The Yardney with its lighter power pack will reach 55 mph and run for 77 miles at 35 mph on a single charge. In 1904, the Krieger Electric covered 50 miles on one charge of its 20 lead-acid batteries at 15 mph over an "average" road of the day. Later models would go 75 miles at 25 mph, a performance not far from the present-day converted Renaults.

The Detroit Electric was built in 1938, looked like a Packard or a Chrysler, but in performance and range it was no better than electrics built more than a decade earlier.

About 1900 to 1910, electric automobiles were widely used for both public and private transportation, especially in larger cities. New York City had a fleet of electric taxicabs, and cab stands all over town sported battery chargers for use between fares. Electric trucks were also in use and, while batteries needed frequent charging and had to be replaced every 3 or 4 years, both cars and trucks seemed almost indestructible.

Curtis Publishing Company, for example, ran a fleet of electric trucks in Philadelphia until 1963. Several of the trucks put into service around 1910 are still in use within the grounds of the company's printing plant at Sharon Hills, Pennsylvania. United Parcel Service in New York still has some 20 electric delivery trucks in operation.

Even in its heyday, the electric car could not meet the requirements of those who needed it most. Most Americans lived and worked on farms, but the electric was essentially a town car.

Electric car design expired with the Detroit Electric in the late 1930s and has not been revived. If an electric car were to be redesigned, batteries could provide performance sufficient for much of today's driving.

GM executives have said that, while it is technically possible to build an experimental electric car as GM has done, the cost would be prohibitive for an electric that would provide range and performance comparable with the gasoline-powered car, a requirement GM believes essential to an electric car's success.

Like the Yardney, GM's prototype Electrovair, a Corvair body with a chassis modified for electric drive, is powered by silver-zinc batteries. There are solid-state controls, an ac induction motor, a converter for direct current to alternating current for the ac motor, a voltage regulator, an oil-cooling system, driving controls, and 13 trays of silver-zinc batteries.

The batteries in the Electrovair II cost $15,000 and can be charged only 100 times before they fail. The rest of the Electrovair costs $7,500, making the total cost of the car $22,500. The performance of GM's prototype electric approaches that of a combustion engine car of the same size. The Electrovair accelerates from 0 to 60 mph in 16 sec and moves at 80 mph. The range is limited to between 40 and 80 miles.

The Ford Motor Company and ESB see most of today's driving done in short hops around town and at relatively low speeds.

The Regional Planning Association of the New York Metropolitan area states that the average auto speed over major arteries leading into that city during the rush hours has been clocked at 13 mph. Once in the city, it slows to $8\frac{1}{2}$ mph.

The concept of a "town car," a small, lightweight vehicle that would take up little space, makes the most of the limited power-to-weight ratio batteries and makes it possible to produce a mass-market electric car.

With its sodium-sulfur battery, displayed as a single cell turning a miniature motor, Ford's "city car" is intended as transportation in urban areas and able to clock 40 mph over a range of 130 miles.

General Motors developed a van-type truck driven by fuel cells. Using an external fuel supply, the truck will run without recharging as long as fuel is permitted to enter the reaction chambers.

In the GM Electrovan, 32 fuel-cell modules developed by Union Carbide are fed by tanks of liquid hydrogen and liquid oxygen to form the basic power unit. With a power train that weighs 3,650 lb (compared to the 870-lb power train in a gasoline-powered truck of the same size), the GM fuel-cell-powered van will accelerate to 60 mph in 30 sec, go to 70 mph, and travel 100 to 150 miles on a single charge of its cryogenic tanks.

Electrovan weighs 7,100 lb with more than half of this weight accounted for by the power train. The conventional GM van weighs 3,250 lb and its power train amounts to 25% of the total vehicle weight. The conventional van has more room and more payload.

General Motors views both its Electrovan and Electrovair as strictly experimental vehicles.

On a less profound academic level, the Great Electric Car Race[1] pitted engineering students from Massachusetts Institute of Technology against those from California Institute of Technology. Each group attempted to drive an electric car cross-country from its own campus to the other in the shortest possible elapsed time.

MIT's entry was a converted 1968 Chevrolet Corvair, Tech 1, with an experimental solid-state-controlled, synchronous dc motor, powered by 2,000 lb of nickel-cadmium batteries worth $20,000. A network of recharging stations had been arranged along the route. Cal Tech entered the lists with a modified 1958 Volkswagen Microbus, equipped with a 20-hp traction motor similar to those for forklift trucks and powered by 2,000 lb of lead batteries.

For 27 miles Tech 1 rolled smoothly down the highway; then the car slowed to a halt. Battery emf was almost nonexistent. The 75-ah capacity of the Ni-Cd delivered one-third that amount. Tech 1 spent nearly 200 miles in tow covering the first nine charging stations.

Car voltages started at 103 v, dropped rapidly to 96 v and slowly slid to 60 v. Simultaneously battery temperature rose excessively. Control was by additions of ice. The car was readied for a "supercharge" for nearly 5 hr. The car breezed through the 36 miles to the next charging station at 45 mph. Later another 5 hr had to be spent in replacing electrolytes.

However, the converter malfunctioned, and the standard 12-v battery had to be replaced three times and often recharged via jump cables from the support cars. The cooling pump clogged and ice in plastic bags had to be used to cool the motor.

At Elkhart, Indiana, there was a motor fire from a short circuit in the battery pack and the car body where insulation had chafed through. Arcing took place at a terminal connection at the motor. After charging at Tulsa, Oklahoma, there was an explosion of hydrogen from the cells by sparks from the battery cooling fan.

At 150 miles from destination near Newberry Springs, California (population 400), for 4 hr the crew worked with utility employees in a charging attempt. Diodes at the battery charger were blown, a transformer burned up, and a power-line fuse was blown, with a power-line contactor erupting into a blazing fireball. The components of the charging system had been breaking down in the reverse direction from transients through faulty grounding of the transmission line. A defective switch in the third battery section meant some power for 17.5 miles had been available. The car was towed by one of the support cars. A little later, while preparing for a tow, the motor was left engaged in first gear. The motor, revving at 12,000 rpm, blew apart and crashed the electric car into the tow car. The Tech 1 was towed to the finish line.

The Cal Tech Volkswagen provided for regenerative charging but this was seldom used—it required 200 amp to energize the field coils, which seldom resulted in a net gain in power.

The Cal Tech Microbus outraced the support cars for the first 13 miles. At the first charging station, charging was twice the estimated time as a result of battery heating. Grinding up a slope 15 miles from the charging station at Needles, Calif., the car had 90 sec of battery power left. It crept to the hilltop at 10 mph and coasted down to Needles.

There was a generator towed behind a support vehicle and there were a number of unscheduled roadside charges of the battery.

Ten miles east of Seligman, Arizona, there was a shift from third to second speed on a steep grade to start regenerative charging. There was a loud thud. The bus stopped. Motor overspeed (6,000 rpm) had flung out the armature windings, causing the motor to seize and necessitating replacement. Electric Fuel Propulsion of Ferndale, Michigan, flew a new motor to Phoenix. Replacement took nearly 24 hr.

While approaching Clines Corners, New Mexico, battery discharge rate was very high because head winds were buffeting the bulky bus. So the passengers pushed the bus to get it rolling and walked to reduce the load.

At a charging station at Amarillo, Texas, three silicon diodes blew up in the charging circuit. Battery overheating caused charging times of two to three over that estimated. Bags of ice were used to cool the batteries or were added to the electrolyte. The batteries delivered only 100 ah compared with over 200 at the beginning. Many cycles of charge and discharge developed gross mismatch in all the cells and a long, slow charge of 4 to 6 hr duration was needed.

In the last 18 hr, 650 miles were covered to complete the trip to the M.I.T. campus.

The Microbus made the trip in 8 days, 19 hr, 146 min, stopped 59 times for charging, averaged 0.3 mile/kwh, and consumed $50 worth of fuel for the 4,700-lb car and three passengers.

Obviously, combustion engine design, flow-back and -by devices, catalytic after-combustion, antismog devices, and related components are being developed and tested on the highways.

REFERENCE

1. *Machine Design*, September 26, 1968, Penton Publishing Co., Penton Bldg., Cleveland, Ohio, 44113.

chapter 21

Selection of a Battery

Many approaches have been suggested for the selection of a battery. The effect of rate of discharge has been discussed in connection with each battery system. Systems have been compared as to voltages, maximum current drain capacity in ampere hours, watthours per pound, and watthours per cubic inch for the important primary systems in Table 5 and Fig. 17. Dollar costs per killowatthour have also been given.

Optimum performance can be achieved best by meeting critical needs and subordinating other characteristics.

It is necessary to know:

1. Nominal and minimum voltage
2. Current drain and schedule of proposed operation
3. Desired service, at expected temperature—ampere-hour capacity
4. Limitation of size and weight—consideration of premium for small size with high output capacity
5. Environment and storage life—shock and vibration, high acceleration, high altitude
6. Design of battery as a component in a system—type of terminals
7. Economic comparison of batteries of different systems

Small motors, battery-powered, are made in the hundreds of thousands per day by individual manufacturers, and there are multiple manufacturers. Table 7 gives the electrical data of such motors made by Hitachi. Select a satisfactory, commercially available battery to power a motor as in Table 46 and Tables 47 to 49.

In reference to Table 48, 200-hr rate would be on the basis of continuous service 24 hr a day and the battery would have to be recharged in a week or two. At discontinuous or intermittent service the time between charges might be as long as a year.

Howard[1] suggests a method of predicting secondary batteries within ±15% by converting the energy requirements into average volts, average amperes, and total ampere-hours. He adds a design factor which might be an interpretation of battery efficiency. Table 50 gives his basic data at 80°F at a 5-hr rate for 1 lb of cell. From this unit he computes total pounds of cell to carry the total ampere-hour load.

The concept of "1 lb of cell" is purely theoretical. It assumes that the proportion of container, connectors, plates, separators, electrodes, their

TABLE 46   Hitachi RE 14-2280
(from Table 7)

Voltage
    Operating.............................. 1.5 to 3.0
    Nominal............................... 1.5
Current, ma
    No load............................... 200
    At max efficiency...................... 630
    At max output......................... 860
Output, mw
    At 35% efficiency...................... 380
Power to be supplied, mw.................. 1,081
Battery cutoff voltage..................... 0.9

                      Selection (Table 13)

Battery, 2 cells in series, voltage............ 3
Current for max. motor output............. 860+
From Table 13
    NEDA 901, telephone, ASAF cells,
        Voltage, v........................... 3
        Current, ma......................... 1,000
        Dimensions
            Length, in........................ 3.78
            Width, in......................... 2.69
            Height, in........................ 5.81
            Volume, cu in..................... 54.61
            Weight, oz........................ 44
Military designation....................... BA 225IJ

**TABLE 47 Hitachi Re 56L 30110
(from Table 7)**

Voltage
    Operating.......................... 3–6   6
    Nominal.......................... 3
Current, ma
    No load......................... 100
    At max. efficiency................. 580
    At max. output................... 1,500   1,500
Output, mw
    At 45% efficiency................. 2,500   5,500

            Selection (Table 17)

Battery, v............................... 6
Current, ma............................. 1,500
NEDA, 907, 4 × 6 cells:
    Dimensions
        Length, in.......................... 10.44
        Width, in........................... 2.72
        Height, in.......................... 7.22
        Volume, cu in...................... 199
        Weight, oz......................... 148
    Competitive.......................... BA 249IJ
        for an acceptable time period.......... 7D 8907
                                             EX 5590–7
More conservative..................... NEDA 912

size, construction, and composition is the same irrespective of the size and the type of battery. Obviously this cannot be so.

An actual 1 lb of battery or a complete battery weighing 1 lb has a 20-hr capacity of 5.8 ah or 0.29 amp/hr, or a 12-hr capacity of almost 5 ah. This data is from Table 34, with reference to ESB battery ER-6-2 or military BB215/U.

From the same table battery ERH 25-2 or BB210/U weighs 3 lb, has a 21-ah capacity at the 20-hr rate or a little more than 1 amp/hr, and 1.8 amp/hr at the 12-hr rate. This has a capacity of 21 ah or, at a 3-lb rate, 7 ah/lb.

A sophisticated 24-v aircraft battery is ESB 24-35 from Table 34. This has a 20-hr rate capacity of 140 ah or 7 amp/hr. The 12-hr rate is nearly 12 amp/hr. The weight is 82 lb, or 1.7 ah/lb at the 12-hr rate.

Electrical size is often expressed as capacity at the 20-hr rate of discharge at 80°F. If a battery can be discharged at 5 amp for 20 hr before the discharge voltage falls off, the unit is rated at 100 ah.

From Table 36, a 12-v Delco 839 ordnance battery has a 20-hr capacity of 200 ah and weighs 156 lb; its capacity is approximately 1.3 ah/lb.

TABLE 48 Hitachi FM 36K 08700
(from Table 7—12-v lead-type car
battery specified)

Voltage.................................. 12
Current, ma
  No load............................. 15
  At max. efficiency.................. 50
  At max. output..................... 95
Output, mw
  At 38% efficiency................... 420
Power to be supplied, mw............. 1,150
Battery, 12v and 100 ma

        Selection (Table 36)

Smallest available is AABM—Group 1—21NL
Voltage.................................. 12
Current, 20-hr rating, ah.............. 24
  100-hr rating, ah..................... 24
  Amp/hr for 100 hr.................... 1
Dimensions
  Length, in........................... 9.0
  Width, in............................ 4.9
  Height, in........................... 6.5
  Volume, cu in....................... 286.7
  Weight, lb........................... 25
Competitive units
  AABM 53, SAE 14M2, Delco G49, G55, G59,
  and G65; SAE 17M2, 17MJ, 9MJ3, and 4NE will
  be ample on a 200-ah capacity rating.

From Table 35, AABM heavy-duty commercial 3T has a 20-hr rating of 160 ah and weighs about 60 lb; its capacity is approximately 2.67 ah/lb.

Summing this up, Table 51 gives the data.

Howard gives as an example a motor-driven truck designed for 12 hr actual use under load conditions of Fig. 4. One day is allowed for recharging. Average voltage is 24 v, ambient temperature is 40°F. Graphical integration of the lead curve is 790 ah, average current is 65.8. Twelve-hour rate is low discharge. From his 1-lb, 2-v cell assumption Howard has 13 cells in series for 24 v. From his equations and corrections he obtains a weight of 2,240 lb, with corrections to 2,580 lb.

Calculating the ampere-hour–voltage load and watts, 65.8 amp at 24 v is 1,564 w/hr; this is 18,768 w for 12 hr. If battery ER-6-2 were employed and series connections made, the weight would be 1,877 lb. If battery ERH 25-2 were employed the weight would be 1,340 lb, to which weights of connections would have to be added.

If battery ESB 24-35 were employed, the weight would be 469 lb for this sophisticated high-cost battery.

**TABLE 49   Hitachi RE 34-4535
(from Table 7, nickel-cadmium
battery specified)**

Voltage	1.25–2.50
Current, ma	
No load	830
At max. efficiency	2,500
At max. output	4,150
Output, mw	
At 28.5% efficiency	1,460
Power to be supplied, mw	5,120
At 1.25 v, ma	4,000+
At 2.50 v, ma	2,000+

Selection

Difficult, if service is heavy or continuous.

From Table 40, 2K90 cells in series would give 2,000 ma, but only for 1 hr on continuous service. 2K95 in series would give 2,000 ma for 1.5 to 3 hr. 2K100 (AISI) in series would probably satisfy discontinuous or intermittent service.

A conservative selection would be 4K100, with two sets of cells in parallel and each pair in series. See Table 40 for detailed dimensions.

TABLE 50   Idealized Performance of Hypothetical 1-lb Cells

Cell type (common name)	Zinc-carbon	Manganese-alkaline	Mercury	Silver oxide–zinc (button)
Typical size range, lb/cell	0.003–2.1	0.02–0.5	0.001–0.4	0.004–0.005
Idealized performance, 1-lb cell, 70°F, continuous 1,000-hr rate, 1-v cutoff:				
Open-circuit voltage	1.5	1.5	1.35	1.6
Average voltage to cutoff	1.25	1.3	1.3	1.5
Energy, wh/lb	36	45	52	50
Capacity, ah/lb to 1-v cutoff	30	35	40	33
Specific volume, cu in./lb (complete cell and case)	15	10	7.5	6
Max. current during short circuit, amp, for 1-lb cell	30	20–60	8–18	40

Batteries and Energy Systems

TABLE 51 Watthours per pound of battery

Battery	Table reference	v	ah/lb	wh/lb
ESB ER-6-2	Table 34	2	5	10
ERH 25-2	Table 34	2	7	14
ESB 24-35	Table 34	24	1.7	40.8
Delco 839 ordnance	Table 36	12	1.3	15.6
AABM heavy duty 37	Table 35	6	2.67	16.02

If battery Delco 839 Ordnance were used, the weight would be about 1,200 pounds, and if the choice were AABM 37, about 1,170 pounds.

Table 52 sums up the figures.

Howard's figures are not within $\pm 15\%$, and his calculations omit so many variables and commercial factors that they become very misleading. When applied to actual commercial batteries born in the fire of intensive competition for markets, a much more meaningful comparison would be dollars per watthour or dollars per pound of battery for the same watthours. Such figures have been added to Table 52. The solution to the problem is obviously a production line assembly product, tested in competitive service. The choice becomes anything but the solution Howard proposes, even if he were given the tremendous advantage of assuming very low cost per pound.

F. P. Yeaple[2] also proposed a method based on a theoretical pound cell and used the data in Table 50.

As an example, Yeaple gives one 100-ohm light bulb at 10 v dc, 2 hr per day for 30 days, 30°F ambient. Current 0.1 amp, power 1 w, total 60 wh.

Ideal nominal capacity M-C is 30 ah/lb. From tables and battery efficiency this becomes 19.2 ah/lb. Yeaple uses a temperature correction of doubling cell capacity for each 18°F rise in temperature and lowering

TABLE 52 Economics

Battery	Weight of battery, lb	Cost, dollars		
		Per lb	Total	Per wh
Howard	2,558	0.80	2,046.4	0.109
ERH 25-2	1,340	1.50	2,010.0	0.107
ESB 24-35	469	3.50	1,631.5	0.087
Delco 839	1,200	1.00	1,200.0	0.064
AABM 37	1,170	1.10	1,287.0	0.069

it for each 18°F fall, with 70°F as reference. From curves he now has 14.4 ah/lb. The cell weight becomes 0.458 lb and volume 6.87 cu in. For cells in series (eight) total weight is 3.664 lb and nominal volume 55 cu in.

This is a nominal 12-v, 100-ma battery at 30°F. From Table 17 NEDA 918 is a readily available battery, 12 v, 250 ma, consisting of eight F cells, weighing 3 lb 6 oz with a volume of 69.5 cu in., and it will do the job. As for Eveready 732, the capacity is ample for the service and the battery is off the shelf.

In another example a highway flasher with ambient temperature 30 to 70°F, which is operated 24 hr a day, 7 days a week, for 25 days, and whose resistance is 15 ohms, time on 0.1 sec/sec (10% of the time), cutoff potential 1.1 v (average 1.3 v), and average current 0.087 amp, has a total service time of 60 hr and a capacity 5.22 ah.

Checking Table 13, we find NEDA 11 (two F cells), or NEDA 13 (one D), NEDA 18 (two D), NEDA 20 (one G) or NEDA 23 (one F) will readily do the job, so the solution has to involve economic acceptability, suitability of connections for easy use, as well as ready availability.

REFERENCES

1. *Product Engineering*, October 26, 1964.
2. *Product Engineering*, March 18, 1965.

Index

Index

Adams, G. B., 185
Adams, J., 194
Adams cuprous chloride battery, 4, 6
Allis-Chalmers, 69, 77
American Association of Battery
  Manufacturers (AABM), 6, 126
American Chemical Society (ACS), 6
American Institute of Chemical
  Engineers, 6
American National Standards
  Institute (ANSI), 6, 36, 81, 90
American Society of Testing and
  Materials (ASTM), 6
Ampere defined, 17
Anderson, L. B., 185
André, H. G., 85, 87
Angus, John C., 184
Aquacels, 100
Aqualites, 100
Arendt, M., 4, 6
Atomic Energy Commission (AEC), 198

Batteries:
  lead storage, tables, 128–140
  selection of (*see* Selection of batteries)
  (*See also* Cells)
Berger, C., 184
Betachrons, 199
Bottger, J., 4
Bradd, 24
Burgess Battery Company, 56, 81

Cairns, E. J., 74
Cameron, G. L., 144
Carlisle, 4
Cells:
  air, 59–63
    Le Carbone, 61
      table, 62
    Union Carbide, illus., 60
  alkaline, 28, 54–58
  alkaline-manganese dioxide, 160

218   Batteries and Energy Systems

Cells, alkaline-manganese dioxide (*Cont.*):
    characteristics of, 55
    charging, 161
    dimensions of, 103
    Lalande, 103
    sizes of, 161
  alkaline secondary, 155–160
    characteristics of, table, 161
    construction of, 156
    electrolyte, 157–159
    reactions, 156
    temperature effects, 159
  Bunsen, 4, 102
  cadmium-silver oxide, 96
  Carhart-Clark, 21
  charge-retaining, 148
  Clark, 19–21
  counter, 146–147
  cuprous chloride, 97, 100
  Daniell, 4, 5, 21, 26, 102, 103
  De la Rue, 21
  dry, 5, 33–53
    applications for, 43, 45
    capacities of, 40
    characteristics of, tables, 45–51
    electrolyte, 34, 37, 42
    magnesium, 52
    manganese dioxide in, 34
    manufacture of, 33
    marketing of, 45
    operation, 44
    reactions, 35
    recharging, 52
    service life of, 39
    sizes of, 36
    standard tests, table, 41
  dry-charged, 142
  Edison, 104, 107, 155
    patents for, 6
  Eppley Laboratory, 20, 21
  Féry, 102
  fuel, 26, 32, 64–77
    Allis-Chalmers, 69
      diagrams, 70–72
    efficiencies, 68, 69
    fuels for, 68
    hydrocarbon, 74
    reactions, 65
    storage unit, 186
    systems, tables, 75–76

Cells, fuel (*Cont.*):
    Union Carbide, 65
      illus., 67
  Gassner, 4, 5, 33
  Gouy, 21
  Grove, 102
  half, 16
  Helmholtz, 21
  Hubbell, 155
  intermetallic, 185
  iron-air, 63
  Ironclad, 117
  isotope, 197
  Lalande, 27, 103, 104
  lead, 107, 109
    applications for, tables, 128–140
    characteristics of, 126
    counter, 146
    dry, 125
    dry-charged, 120, 142
    formation of, 115
    manufacture of, 111, 122
    pasted plates, 116
    prices of, 141–142
    production of, 151
    raw materials, 151
    sizes of, 126–140
    temperature effect, 124
    theory, 110
  lead-calcium, 146, 147
  lead secondary, 110–154
  Le Carbone, 61
  Leclanché, 3–5, 33, 103, 105
  lithium, 191–195
  magnesium, 28, 52, 98, 100
  mercury, 28, 78–84
    construction of, 80
    dry, table, 82
    RCA, 81
    service at various temperatures, table, 83
    sizes of, 79–80
  Muirhead, 22
    illus., 23
  nickel-cadmium, 163–175
    button, 169, 170
    capacity of, 169
    charging, 167
    construction of, 168, 170
    electrolyte, 167
    reactions, 164

Cells, nickel-cadmium (*Cont.*):
  sealed, 168
    table, 173
  sintered plates, 171
  sizes of, 173
  voltage, 166, 172
 nuclear, 198
  characteristics of, table, 198
 obsolete, 102
 perchloric, 103, 105, 106
 photovoltaic, 190
 Planté, 3, 4, 107, 112, 146
 Poggendorf, 102
 R.M. mercury, 28, 78
 Ruben, 4, 78
 silver, 28, 85
  applications for, 90
  button, table, 91
  manufacture of, 86
  performance of, 86, 94
  reactions, 85
  sizes of, tables, 89, 92–93
  systems, 87
 silver-cadmium, 174–175
  table, 174
 silver chloride, 97, 98
 sizes of (*see* Sizes of cells)
 solar, 188, 189
 standard, 16
 thermal, 195
 voltaic, 2
 wafer-type, 38
 water-activated, 97
  applications for, 100
    table, 99
  cuprous chloride, 97, 100
  silver chloride, 99
  sizes of, 100
 Weston, 20, 21, 23
 Weston-Clark, 21
 Wallaston, 2
 zinc-silver oxide, 96
 (*See also* Batteries)
Charging, 176
 circuits, 178
 slope characteristics, 179
Chemelectrics, 26
Chronology, battery, 1, 4, 5
Cohn, Ernst M., 30
Coleman, J. H., 197
Compton, K. G., 145

Copeland, L. C., 35
Coulomb defined, 18
Couples, 12
Craig, D. N., 105, 122
Cranking power, 143, 145
Crown of cups, 2
Cunningham, W. A., 148
Current density, 17
Curtis Publishing Company, 203

Daniell, J. F., 3, 4
Daniell cell (*see* Cells, Daniell)
Davy, Sir Humphrey, 2, 4
Denison, I. A., 85
Depolarizers, 27
Dyne defined, 1, 8

Eagle-Picher, 61, 87
Edison, Thomas A., 3, 4, 104, 155, 201, 202
Edison cell (*see* Cells, Edison)
Eiche, 24
Electric cars, 201–206
 Detroit Electric, 203
 Electrovair, 204
 Electrovan, 204
 Great Electric Car Race, 205–206
 Henney Kilowatt, 203
 Krieger Electric, 203
 Yardney, 203
Electric trucks, 203
Electrochemical systems, regenerative, 183–186
Electrode potentials, table, 10
Eliason, R., 194
Elkington, J., 4
ESB, Incorporated, 126, 204
 Wisco Division, 151
Exide Ironclad battery, 117

Faraday, Michael, 2, 4
Faure, 4
Findl, E., 186
Fischer, A. K., 185
Ford Motor Company, 204

Galvani, Luigi, 2
Galvanic concept, 2

Gassner, Carl, 4, 5, 33
Gassner cell (*see* Cells, Gassner)
General Electric, 63, 77
General Motors, 204, 205
Gibbs-Helmholtz equation, 11
Gladstone, 4
Globe-Union cast-on-strap machine, 122
  illus., 121
Goldberg, 125
Gould Marathon Battery Company, 90
Gray, Willard F. M., 1, 2
Greenberg, S. A., 185
Griffith, F. S., 35
Groce, I. J., 184

Half-reactions, 12
Hamer, W. J., 148
Hamm, A. A., 145
Hare, Robert, 4
Haring, H. E., 145
Harned, H. S., 148
Harner, H. R., 148
Hatfield, J. E., 148
Heinrich, R. R., 193
Heise, G. W., 59
Herédy, L. A., 184
Hertz (Hz), 24
Hetaerolite, 35
Howard, F., 208, 210, 212

Intermetallic cells, 185
International Electrochemical
  Commission, 6
International Electrotechnical
  Commission, 36, 81
Isotopes, 197
Iverson, M. L., 184

Jacobi, 4
Jacques, W. W., 73, 74
Jirsa, F., 85
Johnson, C. E., 193
Jordan, 4
Joule defined, 18
Jungner, E. W., 85, 155

Kennedy, J., 194
Klein, M., 186
König, Wilhelm, 2

Lead compounds, 151–153
Leclanché, Georges, 3
Leclanché cell (*see* Cells, Leclanché)
Leesona Moos zinc-air primary battery,
  61, 63
Ley, Willy, 2
Liebhafsky, H. A., 74, 183
Life, indicated battery, table, 151
Linden, 6
Litharge, 153

McCully, C. R., 185
McMurdo Instrument Company
  Limited, 100
"Mag-Air," 63
Manchester plates, 113
Matsen, J. M., 186
Morrison, W., 85
Morse, E. M., 86
Motors, battery-operated, table, 31

NASA's fuel-cell program, 30, 77
National Bureau of Standards (NBS),
  6, 17, 36
National Electronic Distributors
  Association (NEDA), 6, 81, 90
Nernst, Walter, 8, 9
Newton defined, 18
Niaudet, 4
Nicholson, S. B., 185
Nicholson, William, 4

Obsolete systems, 102
Ohm, Georg Simon, 4
Ohm defined, 18, 19
Oldenkamp, R. D., 184
Oxidation-reduction potentials,
  table, 14–15

Passivity, avoidance of, 27
Patent literature, 6
Patterson, Moos Research, 198
Pettit, C. W., 144
Phipps, G. S., 145
Photoelectric unit, 189
Photovoltaic unit, 190
Planté, Gaston, 3, 4, 107, 202

Planté cells (*see* Cells, Planté)
Plates, Planté, 112, 146
Pratt-Whitney, 77
Project NOMAD, 151

Ramsey, W. E., 36
Rayleigh, Lord John William Strutt, 20
Raypaks, 198
RCA mercury cell, 81
Recharging of dry cells, 52
Recht, H. L., 184
Reed, J., 125
Reference electrodes, 11
Regenerative systems, 183
Rescue lights, 55
Reversible systems, 107
Robinson, 4
Rowls, G. A., 144
Ruben, Samuel, 4, 6
Ruben cell (*see* Cells, Ruben)
Russell, 4
Rymarz, T. M., 185

Schrodt, J. P., 105
Schumacher, E. A., 59
Schumacher, E. E , 145
Selection of batteries, 207–213
   for continuous service, 211
   economics of, 212
   for highway flasher, 213
   for small motors, 208–211
Sellon, 4
Single electrode potential, 9, 10
Sizes of cells:
   of alkaline units, 161
   of dry, 36, 37
   of fuel, 75
   of lead, 126–140
   of magnesium, 97
   of mercury, 79, 80
   of nickel-cadmium, 173
   of secondary, 126–140
   of silver, 89, 92

Snyder, 122
Society of Automotive Engineers (SAE), 126
Spencer, 4
Strier, M. P., 184

Thomas, U. B., 145
TRAC (Thermally Regenerative Alloy Cell) systems, 184
Tribe, 4
Tubular forms, 114

Ulrich, G. D., 184
Union Carbide:
   air cells, 60
   fuel cells, 65
United Delco, 146
United Parcel Service, 203
United States Government, 4

Vinal, G. W., 6, 105, 122
Volt defined, 19
Volta, Count Allessandro, 2, 4
Voltaic pile, 2

Wafer-type cells, 38
Weston, Edward, 3, 20
Weston Instruments, Incorporated, 23, 24
White, J. C., 105
Wilke, Milton E., 36
Woolrich, 4
Wright, 4

Yardney Electric Company, 61, 203
Yeager, E., 77
Yeaple, F. P., 212
Young, G. J., 6